GIS for SCIENCE

APPLYING MAPPING AND SPATIAL ANALYTICS

Volume 2

DAWN J. WRIGHT AND CHRISTIAN HARDER, EDITORS

D1504901

Esri Press | Redlands, California

Library of Congress Control Number: 2020945031
ISBN: 9781589485877

Christian Harder and Dawn J. Wright, eds.; *GIS for Science: Applying Mapping and Spatial Analytics, Volume 2*; DOI: https://doi.org/10.17128/9781589485877

Attribution list available at GISforScience.com. All Johns Hopkins University COVID-19 Dashboard images credited to Johns Hopkins Center for Systems Science and Engineering.

Developmental editing by Mark Henry; Technology Showcase editing by Keith Mann; Layout design by Steve Pablo; Cover design and map by John Nelson.

On the cover: The Spilhaus World Ocean Map depicts the oceans of our planet as a single intermingled body of water—which they actually are. This map displays the geomorphology of the seafloor as tectonic activity splits the crust along volcanic ridges and folds older rock beneath neighboring plates along deepsea trenches. Learn more about the unique qualities and interesting story behind the modern GIS incarnation of the Spilhaus map in the Technology Showcase on page 182.

See this book come alive at
GISforScience.com

PRAISE FOR *GIS FOR SCIENCE*, VOLUMES 1 AND 2

"*GIS for Science* illustrates with stunning graphics, colorful imagery, and accessible text how The Science of Where® can be used to visualize and study our planet, to ultimately conserve and protect it. Accompanied by a web-based resource of tools and apps, professional and community scientists alike can explore Planet Earth, from the poles to the ocean, in 2D to 3D, and discover rich data layers integrated in space and stunningly displayed. The perfect balance of theoretical explanation and practical applications, from modeling bird migration to unearthing archeological sites, this book is a must-have resource for understanding our planet and its response to a changing climate."

—Shannon Bennett, chief of science, California Academy of Sciences

"An amazing book with spectacular images of land and sea—maps that communicate how our planet is structured and changes. Examples are drawn from volcanoes, glaciers, tsunami, climate, rivers, landscapes, and oceans, and the state-of-the-art survey methods summarized. Recommended reading for those interested in the environment and its geography."

—Mark Costello, Institute of Marine Science, University of Auckland

"A beautifully illustrated and informative book that will appeal to the expert and lay person alike. By exploiting modern technologies, the reader not only understands the role GIS plays in modern scientific research, but also experiences it through the online story maps that give a window on the world for everyone."

—Helen Glaves, senior data scientist, EGU McHarg Medalist, British Geological Survey

"*GIS for Science* provides beautifully described and illustrated case studies showing the myriad scientific fields fueled by 'high-tech' mapping information. This book takes us far from the imaginary maps stating 'Here be dragons' to clear-cut, modern geographical information systems that can be used to combat climate change and other real-world issues through open science and citizen engagement."

—Laurie Goodman, co-editor-in-chief, GigaScience

"Stunning visuals and accessible graphics make this book a must-have for GIS users, spatial planners, managers, and anyone interested in Earth science. The case studies are real eye-openers, showing the power of new spatial visualization technologies with applications to the real world."

—Peter Harris, managing director, GRID-Arendal

"What an amazing book! This work explores one of the great frontiers of twenty-first-century science—the use of spatial data and analytical tools to understand our changing planet. In this volume, we see the cutting edge of this exciting field and the scientists and engineers creating these emerging tools. This book will inspire another generation of scientists."

—Jon Foley, executive director, Project Drawdown

"Life in the Anthropocene will require everyone to have a greater understanding of their environment and how they can adapt to ongoing change. This visually compelling book from Wright, Harder, and colleagues beautifully illustrates how geography and spatial analysis will be central and necessary to this new understanding."

—Mark Parsons, director, Data Science Operations, Tetherless World Constellation, Renssalaer Polytechnic University

"This gorgeous book is a joy to read, explore, and also experience online. Perfect to connect the curious minds of all ages with exciting scientific concepts and research that is very approachable."

—Shelley Stall, director of data programs, American Geophysical Union

"As humans we are equipped with a natural ability to extract meaning from visual analysis, and GIS has been paving a technological way in this direction. This book gives a great overview of the power of visualization paired with spatial analysis to give us new insights into complex processes in a spatial context. The examples in the book are carefully chosen to illustrate the principles of this approach."

—Jens Klump, geoscience analytics team leader,
Commonwealth Scientific and Industrial Research Organization (CSIRO), Australia

"*GIS for Science* takes the reader through stunning examples of scientific discovery and revelation. From the examination of natural features to understanding the human impact on diverse environments—in our past, and for our future—this book is a wonderful visual and intellectual primer. A must-read for anyone seeking to understand the power and promise of GIS."

—Roberta Marinelli, dean, College of Earth, Ocean, and Atmospheric Sciences, Oregon State University

"This book powerfully illustrates the myriad ways in which GIS provides insights into the world around us. With compelling case studies and striking images, *GIS for Science* will be a source of inspiration for readers of all backgrounds."

—Mike Mascia, senior vice president, Moore Center for Science, Conservation International

"This volume, *GIS for Science: Applying Mapping and Spatial Analysis*, edited by Dawn Wright and Christian Harder, provides a much-needed addition to the GIS literature. It represents a fascinating blending of clearly explained, critical biophysical and human-social processes—from mapping human settlements to understanding polar ice loss—with current geospatial and visualization tools. The volume, which includes many of the major geospatial research projects of global significance, can be used as an enhancement to both basic GI Science or geography courses. Once started, it is hard to put this terrific volume down."

—Robert McMaster, acting executive vice president and provost, professor of geography, University of Minnesota

"One of the greatest challenges of modern science is to integrate immense quantities of data in order to create rich meaning from raw information. The rigor, artistry, and inspired vision make *GIS for Science: Applying Mapping and Spatial Analytics* a joy to explore. Page after gorgeous page introduces experts and newcomers alike to delicious new ways of truly seeing the world around us."

—Liz Neeley, executive director, The Story Collider

"When you combine maps, powerful images, and science, you combine a sense of adventure, a desire to explore, and a pursuit of knowledge. Wright and Harder have done just that, forging all these things to help us see, and imagine, our world and universe… and more."

—Brooke Smith, director, public engagement, Kavli Foundation

"*GIS for Science* is precisely what we need right now. As the world changes at an increasingly rapid rate, mapping and spatial analytics significantly increase our ability to ask key questions and design new solutions. Dawn Wright and Christian Harder remind us of how powerful these tools are and challenge us to do even more."

—Daniela Raik, senior vice president, Americas Field Division, Conservation International

CONTENTS

Islands come in all shapes, sizes, and types, from tiny rocky outcrops to enormous continental landmasses. The true number of islands distributed in the planet's seas and oceans is still elusive. Recent efforts, bolstered by an abundance of detailed satellite imagery and the sophistication of geographic information systems (GIS), are bringing answers to those questions closer than ever.

With a veritable deluge of new data sources for oceans coming online from satellites, shipboard surveys, and autonomous systems, transforming raw data into meaningful information has emerged as a crucial need for marine industries and management across a broad spectrum of communities. The National Oceanic and Atmospheric Administration and Bureau of Ocean Energy Management have successfully deployed an advanced geographic information systems platform to unleash the power of spatial analytics to unlock ocean intelligence.

Since their invention in the 1950s, plastics have had an alarming and highly visible impact on the world's oceans that humanity certainly never anticipated. Modern scientific detectives are turning to big data and advanced GIS software to understand the major sources of plastic pollution in the world's oceans as a first step to reducing their presence.

The majority of the world's energy and mineral resources are extracted from below ground. Subsurface geologists explore the world below land and sea through the lens of current and historical data. Using GIS and geoscience methods, scientists are redefining our knowledge of the subsurface.

Researchers at Oak Ridge National Laboratory are mapping the global footprints of human activity with unprecedented spatiotemporal resolution. With a global population now approaching 8 billion people, this herculean effort demands advanced machine learning, artificial intelligence, and one of the world's fastest supercomputers.

GIS FOR SCIENCE: A FRAMEWORK AND A PROCESS

by Jack Dangermond, founder and president, Esri
and Dawn J. Wright, chief scientist, Esri

Science—that wonderful endeavor in which someone investigates a question or a problem using reliable, verifiable methods and then broadly shares the result—has always been about increasing our understanding of the world. In the beginning, we applied geographic information systems (GIS) to science—to biology, ecology, economics, or any of the other social sciences. It wasn't until around 1993, when Professor Michael Goodchild coined the term *GIScience,* that the world began to realize that GIS is a science in its own right. Today, we call this The Science of Where®. GIS incorporates sciences such as geology, data science, computer science, statistics, humanities, medicine, decision-support science, and much more. It integrates all these disciplines into a kind of metascience, providing a framework for applying science to almost everything, merging the rigor of the scientific method with the technologies of GIS. The study of where things happen, it turns out, has great relevance.

We live in a world that faces more and more challenges. In 2020, we confronted a global pandemic, the likes of which we have not seen in our lifetimes. (Learn about the early and near real-time response of the GIS community to this world crisis in these next pages).

We also continue to see, hear, and read about such issues as growing population (some would say overpopulation), climate change, loss of nature, loss of biodiversity, social conflicts, urbanization, natural disasters, pollution, and political polarization. We also confront the realities of food, water, and energy shortages, and general overconsumption of resources. Although the pandemic took center stage and an unprecedented response as it spread worldwide starting in early 2020, these other concerns are not trivial for the individuals and organizations working in these fields. We must do everything we can to better understand these crucial issues and form better collaborations to address the challenges.

Our world at the same time is undergoing a massive digital transformation. Science always has been about increasing our understanding of the world. But it is also about using that understanding to enable innovation and transformation. It is about what we can measure, how we analyze things, what predictions we make, how we plan, how we design, how we evaluate, and ultimately, how we weave it all together in a kind of fabric across the planet.

What GIS provides is a language to help us understand and manage inside, between, and among organizations, to positively affect the future of the planet. It is also a framework in which we can compile and organize maps, data, and applications. We can visualize and analyze the relationships and patterns among our datasets, perform predictive analytics, design and plan with the data, and ultimately transform our thinking into action to create a more sustainable future. This technology also delivers a new way to empower people to easily use spatial information. As Richard Saul Wurman has said, "Understanding precedes action." Esri is driven by the idea that GIS as a technology is the best way to address the immense challenges of today and the future.

> Science itself is driven by the organic human instinct to dream, to discover, to understand, to create, and to help each other in times of great need.

This book is full of examples that show how GIS advances rigorous scientific research. It shows how many science-based organizations use ArcGIS as a comprehensive geospatial platform to support spatial analysis and visualization, open data distribution, and communication. In some cases, we use this research to preserve and restore iconic pieces of nature—revered and sacred places worthy of being set aside for future generations. These places belong to nature, and they also belong to science.

As scientists, the discipline of the scientific process is the central organizing principle of our work. But science itself is also driven by the organic human instinct to dream, to discover, to understand, to create, and to help each other in times of great need. The Science of Where is a concept that brings these impulses together as we seek to support and transform the world through maps and analytics, connecting everyone, everywhere, every day through science. At Esri, we are encouraged and frankly humbled by the often heroic work of the GIS community.

INTRODUCTION BY THE EDITORS

Dawn J. Wright and Christian Harder, Esri

This book is about science and the scientists who use GIS technology in their work. This contributed volume is for professional scientists, the swelling ranks of citizen scientists, and anyone interested in science and geography. Our world, now two decades into the twenty-first century, seems to be entering a crucial time in history in which humanity still can create a sustainable future and a livable environment for all life on the planet. But if we look critically at the facts, no informed observer can refute the reality that the current downward trajectory does not bode well.

As work on the book was well under way, we saw a great shift as the GIS community pivoted almost overnight in response to the outbreak of a new coronavirus. The stories we present remain just as relevant, but we also wanted to provide some context about the initial response of the GIS community to this global crisis. Our first objective in assembling volume 2 of this work was to select relevant and interesting stories about the state of the planet in 2020. As such, this is indeed a storybook. It is not an atlas, it is not a research monograph. Even so, we still looked for a cross section of sciences and scientists studying a wide range of problems.

GIS has found its way into virtually all the sciences, but the reader will notice that earth and atmospheric sciences are especially well represented. Web GIS patterns and a simultaneous explosion of earth-observation sensors fuel this growth. Between all the satellites, aircraft, drones, and myriad ground-based and tracking sensors, the science community is now awash in data. Well-integrated GIS solutions integrate all this big data into a common operating platform—a digital, high-resolution, multiscale, multispectral model of our world.

Despite all these advances, science is under attack on many fronts. From fake news to political pressure, science is too often used as a political tool at a time when level-headed, objective scientific thinking is needed. We are convinced that GIS offers a unique platform for scientists to elevate their work above the fray. We invite you to read these stories in any order; the common thread is that all this work happens at the intersection of GIS and science. As you read through these stories, you'll see that the use of GIS as a cross-cutting, enabling technology is limited only by our imaginations.

In some cases, like the fascinating work of the Virginia Commonwealth University using drones and artificial intelligence to count fish, GIS and spatial analysis are at the core of the science. These innovations in science could only happen in the context of an advanced GIS. In other cases, like the story of the NASA Disasters team and its mission to publish the astonishing volume of imagery, GIS embeds itself in the science but is still mission-critical in terms of how the team turns data into information products for public use. GIS also serves as a vital storytelling platform that brings critical research to stakeholders in their communities.

How the book and website work together

It's impossible to describe the full breadth and scope of what GIS means for science and scientists without showing digital examples. So we have created a companion and complement to this book online. You can access it here:

GISforScience.com

This unique website, comprising collections of ArcGIS StoryMaps, apps, and digital maps, brings the real-world examples to life and demonstrates the storytelling power of the ArcGIS® platform. The website also includes links to learning pathways from the Learn ArcGIS site (Learn.ArcGIS.com) and blogs related to the practical use of ArcGIS in each of the case studies.

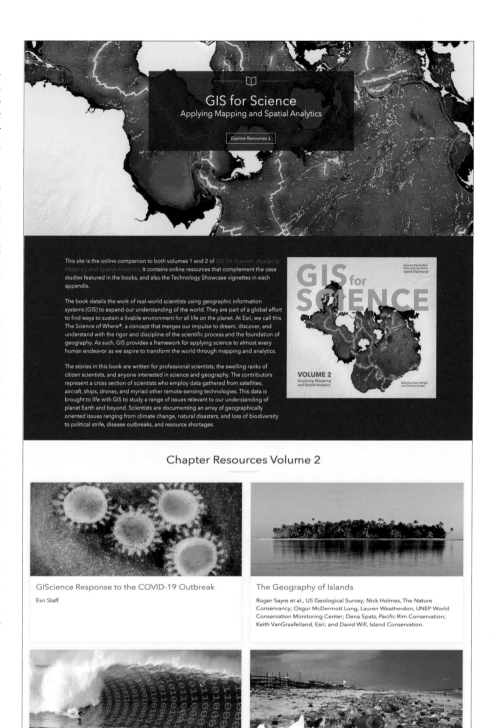

The companion website for this book is in many ways the most important component of the project. Visit GISforScience.com to access more than 100 interactive web maps, apps, story maps, videos, and other digital resources described in the text.

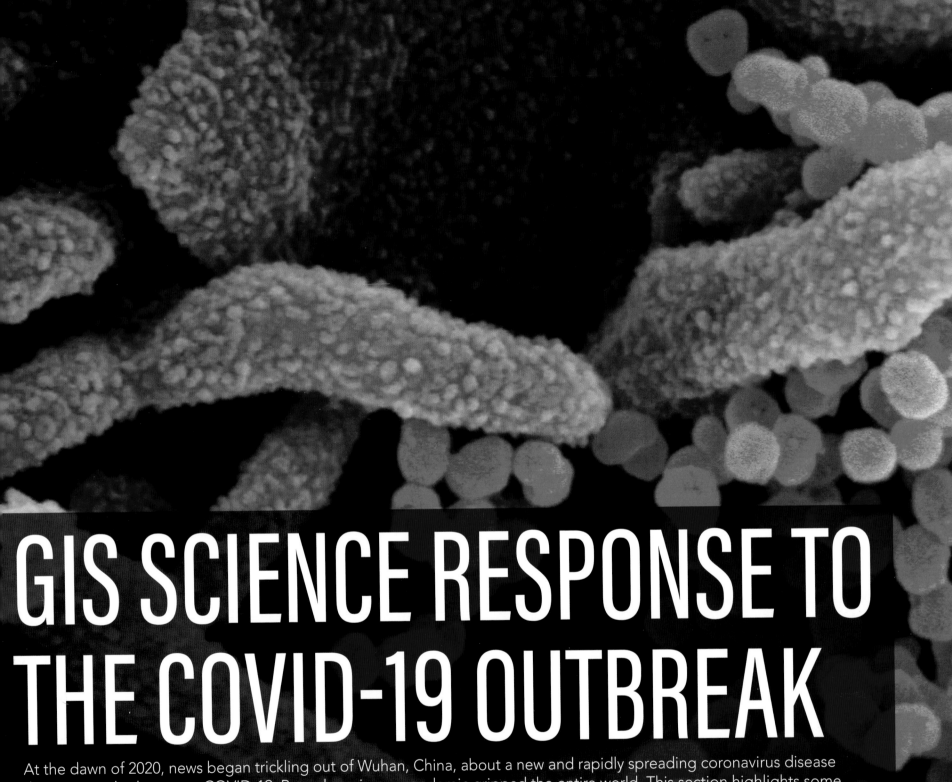

GIS SCIENCE RESPONSE TO THE COVID-19 OUTBREAK

At the dawn of 2020, news began trickling out of Wuhan, China, about a new and rapidly spreading coronavirus disease that came to be known as COVID-19. By early spring, a pandemic gripped the entire world. This section highlights some of the early responses by the global GIS community. Disease and epidemiology are uniquely rooted in place, so spatial analysis and mapping were a natural toolset to deploy.

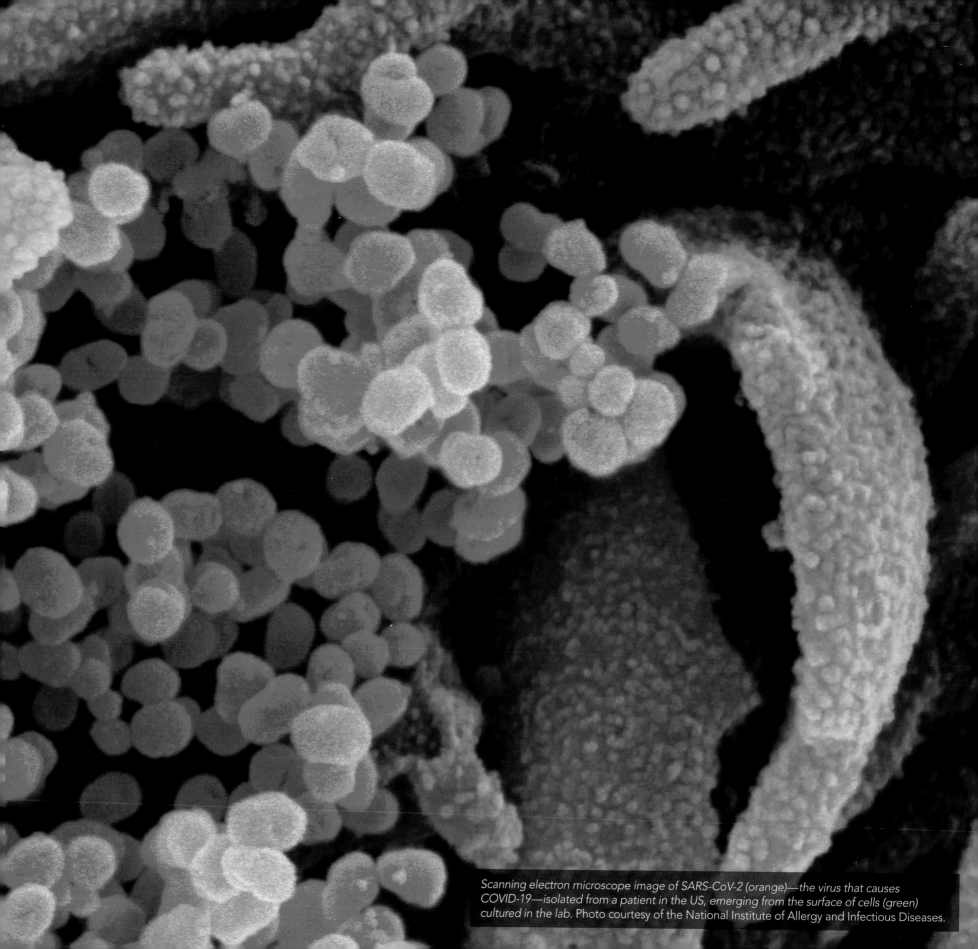

Scanning electron microscope image of SARS-CoV-2 (orange)—the virus that causes COVID-19—isolated from a patient in the US, emerging from the surface of cells (green) cultured in the lab. Photo courtesy of the National Institute of Allergy and Infectious Diseases.

CHRONOLOGY OF A PANDEMIC

To most of the world, the mystery illness seemed far away when the Chinese government reported a new virus to the World Health Organization on the last day of 2019. But by late winter of 2020, the previously unknown virus had swept across the planet, touching the lives of millions of people, causing historic impacts to the world economy, and requiring monumental changes in the ways humans relate to each other. The SARS-CoV-2 virus has presented challenges unlike any that humanity has experienced or seen in our lifetime. The world has witnessed or personally experienced heartbreaking scenes of grief and sometimes despair, frustration, uncertainty, and crisis fatigue. But it also has witnessed inspiring and emotional scenes of hope and resilience from people around the world: Italians singing together from their balconies, children scrawling messages on sidewalks, teachers holding drive-by car parades to cheer up students quarantined in their communities, doctors in scrubs and masks waving and smiling from their emergency rooms, and police officers bringing words of cheer and sometimes even dance routines to the people they serve.

December 31: Chinese authorities inform the world about a mysterious surge in pneumonia cases with no known cause in Wuhan City, Hubei province.

January 9: China reports first death linked to the new coronavirus.

January 24: Japan and the United States each confirm their second COVID-19 cases.

February 2: The first COVID-19 death outside China is reported in the Philippines.

February 27: Brazil confirms its first case of COVID-19, marking the first case in South America. Cases of the virus have now been confirmed on every continent except Antarctica.

February 29: First death in the United States.

March 11: WHO declares a world pandemic.

March 15: The United States officially becomes the country hardest hit by the pandemic, with more than 80,000 confirmed infections and more than 1,000 deaths.

March 24: The Tokyo Olympics are delayed until 2021.

December 2019 — January 2020 — February — March

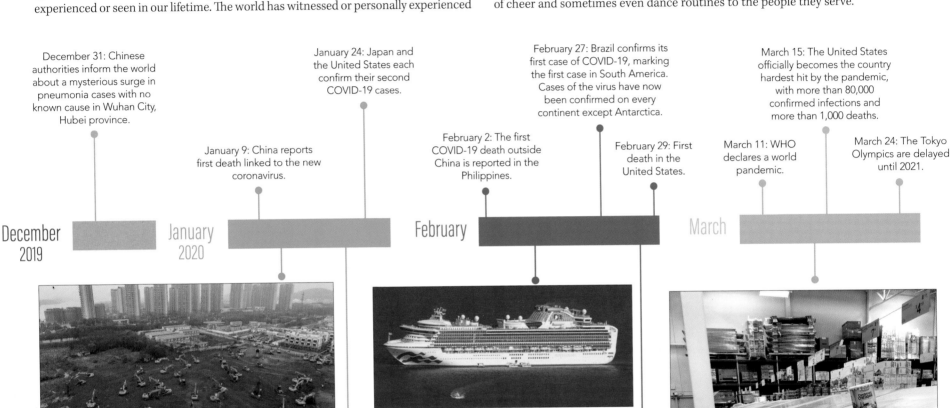

January 24 photo shows rapid construction site of a new hospital in China to treat patients infected by a new virus.

A total of 175 people test positive in early February for COVID-19 on the Diamond Princess cruise ship, quarantined in Japan.

Panic buying in response to the pandemic stripped store shelves nearly bare in Lufkin, Texas.

Residents in Wuhan, China, line up to buy masks in January.

After Italy sees a surge of cases in February, residents find ways to bring cheer to each other; here, performers play the Italian national anthem from an apartment window.

As the crisis unfolded, governments at every level stepped up and began responding. Thousands of organizations, institutions, private businesses, scientists, and researchers began gathering data, attempting to organize the incredible stream of information coming from every country on the planet. As the health providers expanded medical treatment, and public agencies began mandating stay-at-home orders, quarantines, and social distancing protocols, a small army of mapmaking professionals and location-savvy data scientists began feeding data into spatial databases through the technology of geographic information systems (GIS).

Organized geographically in *layers*, raw data can be analyzed in context against other layers and transformed into powerful visualizations. As the outbreak grew, thousands of GIS-powered applications, dashboards, and maps tracked the spread of the virus and informed frontline efforts to fight the disease. One GIS map in particular appeared online early in the crisis. Next, you'll learn how a small team of researchers at the Johns Hopkins University Center for Systems Science and Engineering created the most viral map in history.

April 2: The US Labor Department reports that 6.6 million people applied for unemployment benefits in the last week of March.

April 29: COVID-19 has killed more than 200,000 people worldwide.

May 2: Russia records a one-day record for the country with 9,623 new coronavirus infections. The mayor of Moscow says 2% of the city's population has coronavirus.

May 27: Four months after the first confirmed case, the United States records more than 100,000 deaths from COVID-19.

June 16: After a 24-day span with no infections, New Zealand records new cases.

June 28: Global reported deaths exceed a half million, with total confirmed cases surpassing 10 million.

July 1: The United States reports 55,000 COVID-19 cases in a single day, the highest single-day total in the pandemic's short history.

July 7: President Jair Bolsonaro of Brazil discloses that he has been infected with the virus.

April   May   June   July

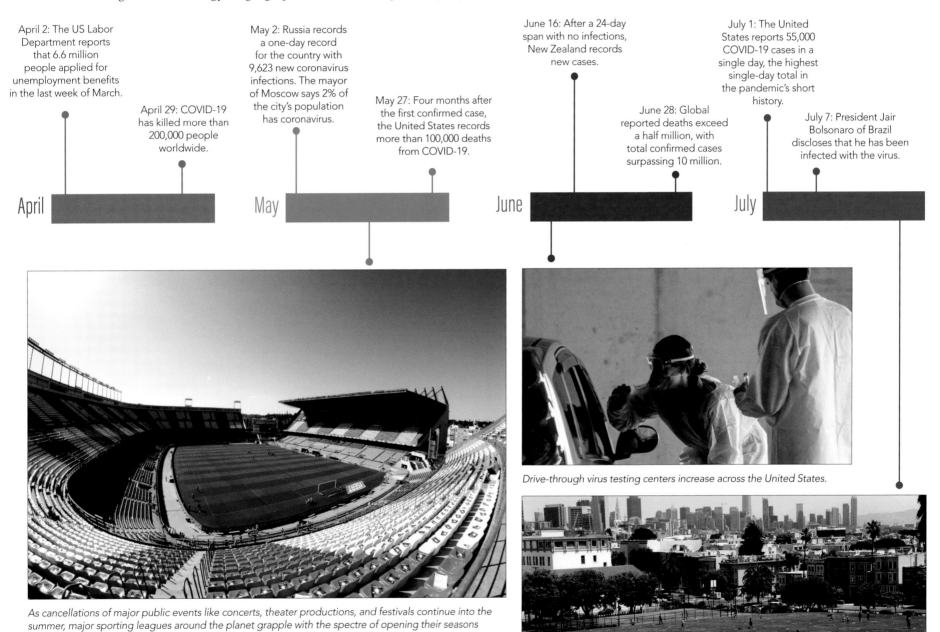

As cancellations of major public events like concerts, theater productions, and festivals continue into the summer, major sporting leagues around the planet grapple with the spectre of opening their seasons without fans in the seats. (Photo by BrudeW/Flickr.)

Drive-through virus testing centers increase across the United States.

As summer temperatures arrive in California, civic leaders take sometimes controversial measures to enforce social distancing.

THE STORY BEHIND THE JHU COVID-19 DASHBOARD

An Interview with Dr. Lauren Gardner

In late January 2020, as the world began to fathom the magnitude of the public health crisis originating in Wuhan, China, a well-designed map began circulating on social media and then quickly found its way onto mainstream media and traditional news sites. The map was actually more of an information dashboard with statistics and charts surrounding the map itself. It drew attention with its dramatic red circles, scaled by size, that indicated the total number of COVID-19 cases and deaths. Perhaps even more important, the daily updates satisfied the desperate need for current and accurate information from a worldwide audience. At the time, the dark-themed application created by the Center for Systems Science and Engineering (CSSE) at Johns Hopkins University seemed to be the most comprehensive and reliable source of information about the spread. Innumerable dashboards to track the virus have since been created (many of which used the underlying JHU data), but the now-famous "JHU dashboard" remains the leading source of consolidated COVID-19 data as of this writing, ultimately generating hundreds of billions of total views.

Snapshot of the JHU COVID-19 dashboard on January 22, a day after its initial launch.

So what is the back story behind the world's most *viral* map in history? Puns aside, the map came about because of an interesting confluence of people at the university who had an idea, a California mapping software company that provided the technology, and the pressing need to fill an urgent information vacuum. When traffic to the dashboard grew exponentially almost overnight, the company, Esri, in Redlands, California, stepped in to help scale it up and handle the extreme traffic as this single app became one of the most viewed maps of the twenty-first century.

The editors of this book and ArcGIS Living Atlas of the World Program Manager Sean Breyer met with Lauren Gardner, associate professor in the Department of Civil and Systems Engineering at Johns Hopkins Whiting School of Engineering who co-created the application, to hear the inside story of the dashboard (the interview has been edited for clarity).

Let's start at the beginning. You were living your normal life and doing epidemiological modeling research at the JHU Department of Engineering when you heard about a new disease outbreak in China, right? So when did this journey actually start?

LG: In early January, I was at one of our normal weekly lab meetings with our PhD students. Ensheng Dong, who goes by Frank, is a first-year PhD student, and his family is in China. We were talking that day about measles and Zika and flu and everything that's not COVID. And then he brought up this new suspected virus in China. He was particularly interested in the outbreak that was happening. This was

beginning to be big news in China at that time. It wasn't even called COVID yet; it was just a few hundred cases around Wuhan. My initial reaction was to say that we're way too busy to follow this new thing, but then pretty quickly, I was like, "Okay, this is actually pretty interesting." This was in the middle of January. Frank was personally really interested in it because of his family, and there was also a lot of uncertainty on the ground in China about what was happening. But it turns out there was some data available in China: daily case data linked to location. So that day we grabbed the data and decided to start collecting it regularly and make it publicly available. The idea was to aggregate all the data, organize it, and share it out. But we also decided we would map it as well. After all, what more of a geographically oriented subject could there be than spread of a transmissible disease?

Frank, who actually interned at Esri for a little while and is really skilled in GIS and spatial modeling, built the first dashboard that night to show the data in a map-focused way, and then we shared it publicly the next morning, on January 21, to be exact. On that first map, I think there was one case in South Korea and one in the US, and it was pretty much just a few hundred cases in China.

I was motivated by the idea of collecting and sharing data for an infectious disease, a novel, new emerging disease, in real time, and sharing it because this is a very data-poor field of epidemiology that tries to understand transmission and risk and spread, and doing that in real time is hard because usually that data just isn't available. I thought, building out a dataset like this that we'd just do as a daily release would be something that, not knowing where it was going to go, may or may not be useful for us and for the community. It took just a week for that thing to take off. So yes, it was good to help the infectious disease research community.

Was there a particular media link or coverage that created the first spike in usage?

LG: Well, I tweeted it on the 22nd, that morning, and that got retweeted and recirculated quite a bit. I could see it was being pretty heavily viewed and retweeted on Twitter because it was trackable, but I'm pretty sure it was circulating on lots of other social media platforms as well; I just wasn't following them because we were busy. But then, all of a sudden, we started to notice the increased traffic. It just kind of snowballed. We actually realized something was happening because Julia Holtzclaw and Sean Breyer from Esri reached out at the end of January and were saying this looks big and important, and you probably need our help to scale it up if the traffic keeps going like this. I was like, "Hi. Yes please," so Esri jumped in and has been supporting this ever since.

Lauren Gardner's January 22 tweet about the launch of the dashboard.

Sean Breyer: So initially when we contacted them, really within the first week, we were seeing some growth, nothing large at that point, but we were starting to see growth, and it was actually set up under a student account; it wasn't even running on a dedicated account of the center (CSSE). It was, "Hey, there's a COVID-19 dashboard app set up by a student, and it's starting to get a lot of hits." I read Lauren's blog that was referenced in the dashboard, and as I was reading through it my first thought was, "I think they're actually updating this thing manually, like once a day, to try and keep up with the numbers. We should probably talk to them and see if we can help."

LG: We were manually doing a daily update of a few different regions, which was not particularly problematic. But then it quickly scaled from a few to dozens to hundreds of reporting sites. And it was in every country, and the number of countries afflicted was growing; yes, it quickly overran our manual processes. Sean, when was the first time we had some infrastructure issues because of demand on the service?

SB: It was mid- to late February when we were starting to see the numbers climb beyond even what ArcGIS Online did on a normal day—just for this one dashboard. So, we took on the challenge of where to host it while Lauren's team focused on the data and map design within the dashboard. Most of the infrastructure changes going on behind the scenes were in our web operations area. We began by isolating their work from everyone else's because most AGOL content is in shared pool environments. So we started to separate it more and more. At the same time, we started automating a lot of the processes. So rather than going to a manual scraping, we helped automate the scraping of the pages that they were using so that they could continue to do daily updates. This worked pretty well until they encountered some of the challenges that come up when you're scraping data from websites rather than receiving structured and validated data. Even one change in the design of a website can mess up everyone who is scraping that data.

How has the workflow evolved over the ensuing months?

LG: It's gone through three phases. There was the first one that was almost fully manual, and that was mostly just me and Frank. And then there was this middle one that lasted for a while where it was our CSSE team and Sean's team working together to semi-automate data collection from one Chinese website that was kind of the premier site for data at the time. We were getting good Chinese data from there, and then they also started reporting data on other countries.

So we used that as our source, but as time went on, the country data for outside of China was experiencing delays, so we were seeing other websites popping up to provide more updated, timely information. So we started trying to do US, Australia, and Canada at the city level from the start, so those were all manual the whole time. And it was starting to be nearly impossible to update every country daily. So that was when we started making this big shift to the third phase, and we expanded the team to include multiple software developers from the JHU Applied Physics Lab. We rearranged tasks and took over all the data curation on the JHU side, and since then have expanded the automation to include dozens of sources that cover countries all over the world, states, cities, counties, so we're just scraping from all sorts of levels. And it's aggregated now to subnational data, and we're still adding sources and building that out all the time. But the first two versions were like that manual one and then the one with Sean's team and us that lasted for a long time. That's really where the dashboard grew out of and got popular, and now it just works a lot better. And Sean's team's got to spend their time and efforts doing the infrastructure management, which is also improved massively, but I don't have any idea what magic they've done! At this point, we've gotten to focus on the data curation, quality,

Ensheng (Frank) Dong, left, and Dr. Lauren Gardner.

Photo by Will Kirk: JHU

robustness, and anomaly detection, and build out that pipeline. So, it's definitely gotten more stable and smoother, but it's been a daily commitment from around 20 people that are still trying to tweak this thing on a regular basis.

What were some of the design challenges?

LG: Frank has a good design eye, and he would make the initial choices as far as colors and symbol sizes and overall arrangement of the dashboard elements. And then I would come in and be picky and regularly make him sit down with me every time we wanted to make changes and try a million different things and resize things. Sean got a lot of emails from us to implement some of these changes. We regularly had to reassess our design choices because the reporting was changing everywhere. For example, we were always having to manually scale the red bubbles into these discrete categories. And that was challenging from the start, because clearly, this is a highly nonlinear phenomenon.

So Lauren, this book is about how scientists use our technology to optimize their science, but the flipside that we often do not hear about is how the science helps Esri to improve our technology and push the boundaries of what we are trying to make in terms of software and services. So, your description here of the challenges of trying to capture a nonlinear event is a very good thing because it's helping us, Esri, to improve.

LG: I do want to highlight that the software is awesome. The software has been amazing, and the support has been great. We were in a unique position where we were dealing with something live, and we had all these ideas and features and layers we'd like to include, but every time we had to have this conversation about adding things, we were constrained because of how popular this thing was. Under normal circumstances, you would just add all these features and layers and keep changing things.

SB: If there were a couple thousand viewers it would have been no big deal at all—no one would have noticed. I wanted to mention, too, that what's unique about our platform, one of the big values that we discovered, is not just the dashboard; it was the underlying hosted service that they built. This service was published as a standalone resource, freely available to anyone who wanted to use it. It ended up being reused by thousands of organizations. So a change in data structure would have messed up thousands of people who were connecting to it to make their own versions of the dashboard that focused on the area of interest, or they were running it to do analysis, so the dashboard became a visual hub for the public, but behind the scenes, the hosted service that was there became the tool, the disaster response and health community's reason to make decisions. So those two were actually being driven by one set of services.

LG: That's a good point. From the start, this whole thing was about open data and open science. So the whole service was open; people could pull the feature layers, and we also deposited all the data that went into the dashboard into GitHub, and that was available. We have to be careful about the data structure because we have all these files there that if we want to add variables or ever change the structure of these files, it affects people all over the world who are pulling those data on a daily basis and utilizing them in their own modeling tools and visualizations. And any time we change anything, it does the same thing to their scripts. So it has to be justifiable to make these structural changes. We can't be selfish about making our own dashboard because every time we want to do something, we have to think about how it affects everybody else that's using it.

That is such a fantastic point. We are learning so much now about culture change and science in terms of opening up our data and our methodologies and our workflows. It used to be when you were a student, you'd do something, and it wouldn't go beyond your professor and maybe eventually a published academic paper. But today, we are getting used to the fact that thousands of people may be depending on our data and workflow, and this is such an apropos example of that.

LG: It's true. I don't think we do any work in my group anymore where you could publish unless all the data you use in your work can be published alongside of it. Today, we share out everything, everything's open. I think that's the only way to do science now. So now you actually have trust issues if you don't make the data available. This COVID dashboard is such a great example of that. A major research interest of mine, something I was doing before this all came about, is about issues of misinformation and disinformation and the growing lack of trust in science and empirical understandings of how the world works. A lot of this is just about being a transparent source of information to the public, so that they can see what's happening, but also for scientists to have open access to the data as well, so that they can build models that are also transparent in terms of where the data's coming from.

Many people have been surprised to learn that this dashboard that has become the de facto "authoritative" source of information about the pandemic would come from a private university as opposed to a government agency. What about that?

LG: I think there's value in it coming from a university. And it's been great that our university has been so supportive and that we didn't have to fight for the right to continue. I was conscious from day 1 of where I would accept funding and support for this to make sure it didn't get branded with any organizations that would take trust away from it. And this was a delicate issue with the federal government because of how politicized this pandemic was from the start and the way they were censoring the CDC and some of the science. So I think that whether or not it should have been done by the government is kind of a separate question, but I do think the fact that it came out of Johns Hopkins, a highly respected institution in public health and medicine, has been a huge benefit. But I will note that it is not in the public health school or the school of medicine; it's in the engineering school, which is also really great!

SB: So that does bring up an interesting question that you and I have talked about a few times as the number of cases grew in the US and we started to collect information.

In trying to get it down to the county level, there were a lot of challenges because all the sources were not just recording stuff differently but using different software products, and the ability to access the underlying data was a challenge. Can you talk a little about that?

LG: Yes, it's been a huge challenge because the reporting criteria and guidelines and structure and the types of things being reported are still constantly changing, but yes, I think maybe what you're getting at, and maybe something that's been an interest from the start, is that we need more systematic and strategic guidelines and processes for reporting, moving forward, that counties and states can follow that all align into a system where it's available in a timely manner. There are, of course, privacy issues, and things have to be aggregated and anonymized when they scale up. But here we were trying to collect data, and there's some three thousand plus counties in the United States, and all the counties report data differently, the counties in a state might report differently than their state, and there are all these inconsistencies at all levels.

And then you have not only this issue of cases and deaths, but also you might get probable cases and probable deaths, and then there's the issue of testing, and so many little things around this that were (and still are) a challenge. We'll see a city or a county in one state that's reporting something, and the state doesn't even report that. And the state says something different about that county than the county says about itself. When we started this, there wasn't a single COVID-dedicated website by anyone, by any government at the state, national, city, or county level. And now almost all of them have it, and we're going through those and trying to go straight to those as our sources to pull data directly into this dashboard. This is, in itself, amazing and one of the remarkable outcomes of all this. I can say with some confidence that people in the world of epidemiology and health science reporting will be much better equipped for the next pandemic.

It is crystal clear that there needs to be a system in place so that when the next thing happens, not five months into it but weeks into it, everyone's already put up their county dashboard, and here's how we report the data, and this data is pulled from this dashboard into this centralized state dashboard, and the state dashboard data can be pulled directly into a centralized US dashboard. And everyone, every county, is reporting the same variables, at the same time period, and things like that. It's complicated, but it can be done. We did it, you know? We have built it, and we're just some engineering professors and students.

What should the GIS community be doing differently in the next pandemic?

LG: GIS people should have the systems in place and connected, and the data provided in some kind of consistent format, so that they can be pulled together into some centralized system that is open and public and accessible and usable. The CDC collects data from states and counties, and they don't share it, which is useless to us. So the public can't see it, so they don't trust it, researchers can't get access to it, so they can't even use it, so I mean from day 1, the stuff we've been doing, even in January, people were using this to help policy makers in China understand what was going on, and it's been used since then for every country as it got hit.

This has been a fascinating conversation. Thanks so much, Lauren, for talking with us.

LG: You're welcome. Thanks to you and thanks to Esri for being so responsive and supportive during this time.

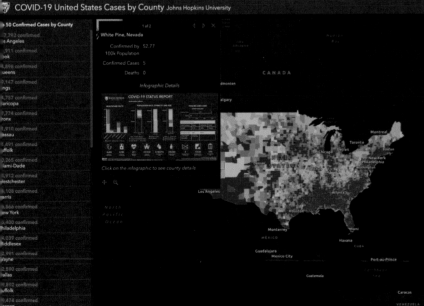

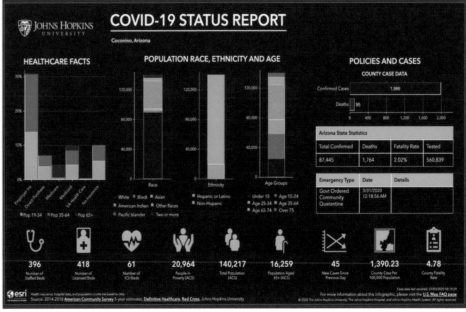

...atistical views of the evolving pandemic as of August 5, 2020, including the cases by country; expanded county-level infographics for all US counties; and a a breakdown of cases by race, ethnicity, and age.

IMPORTANCE OF LOCATION

Esri Chief Medical Officer Este Geraghty

Location information has proven critical to decision-making associated with large outbreaks, and COVID-19 is no different. The US Centers for Disease Control (CDC) considers place to be a basic tenet of a field investigation: Both the *who* and the *when* of disease are relative to and often dependent on the *where*. Epidemiologists quickly turned to GIS science, systems, software (collectively known as GIS), and methods as a needed perspective to understand and track the spread of the virus. The health care community has used maps to understand the spread of disease for a long time, most famously in 1854 when Dr. John Snow connected location and illness with his history-making map of a London cholera outbreak. From disease atlases of the early twentieth century to more recent web mapping of Ebola and Zika infections, health care professionals have long considered mapping, and more recently GIS, a critical tool in tracking and combating contagion.

GIS is critical to answering many infectious disease questions:

- How quickly is the infection spreading, and where is it going?
- Do we have schools in socially vulnerable areas?
- Which neighborhoods are distant from a testing site?
- Do we have communities at a greater risk?
- Which facilities and staff are in harm's way?

What does surveillance data on the number of hospitalizations and deaths suggest about the following questions?

- How are hospital beds and supplies distributed on a regional basis?
- How quickly are local and regional hospital resources depleted?
- Does data help predict where and how fast the pandemic will spread?

The need for location intelligence is acute when an outbreak like COVID-19 quickly spreads from a small geographic location to widespread areas. Public health officials face a major challenge, never before undertaken at this scale, of containing the outbreak through contact tracing and quarantine, which proved to be successful after the new coronavirus was identified in the city of Wuhan, China, in December 2019. For most of the world, however, health officials must evaluate and implement a series of community-level interventions to slow the spread of the illness. Health officials can use location-based information to support multiple, specific community interventions and activities. Common and helpful GIS applications include mapping and data collection apps to track cases, spread, vulnerable populations and places, and the capacity of our systems (like health care) to respond; dashboards for real-time situational awareness; and web apps for keeping the public informed. Health officials may overlay outbreak data with other location-based information, such as public gathering places, schools, health facilities and services, transportation centers, and local population demographics. GIS-supported interventions led to the implementation of many public safeguards, and GIS continues to help monitor their impact in many ways:

- Canceling public events, meetings, and gatherings
- Closing schools, public places, and office buildings
- Restricting use of public transportation systems
- Identifying potential group quarantine and isolation facilities
- Enforcing community or personal quarantines
- Screening airline passengers and assessing airline routes

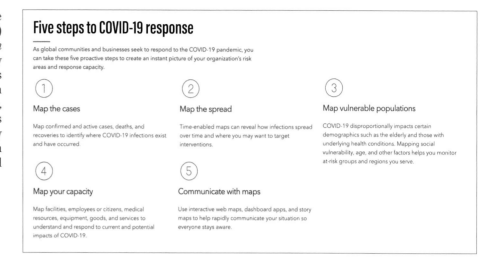

Five steps to COVID-19 response

As global communities and businesses seek to respond to the COVID-19 pandemic, you can take these five proactive steps to create an instant picture of your organization's risk areas and response capacity.

① Map the cases

Map confirmed and active cases, deaths, and recoveries to identify where COVID-19 infections exist and have occurred.

② Map the spread

Time-enabled maps can reveal how infections spread over time and where you may want to target interventions.

③ Map vulnerable populations

COVID-19 disproportionally impacts certain demographics such as the elderly and those with underlying health conditions. Mapping social vulnerability, age, and other factors helps you monitor at-risk groups and regions you serve.

④ Map your capacity

Map facilities, employees or citizens, medical resources, equipment, goods, and services to understand and respond to current and potential impacts of COVID-19.

⑤ Communicate with maps

Use interactive web maps, dashboard apps, and story maps to help rapidly communicate your situation so everyone stays aware.

Early in the crisis, public health officials began screening international airline passengers who completed a standardized health status questionnaire and had their temperatures checked for fever. Passengers stated origin and destination addresses. Subsequently, when a disease cluster was reported in another country, public health officials could better identify how many people traveled from or had visited that same location. Using GIS, public health officials applied the information collected in the questionnaire to estimate exposures and prioritize investigations. A digital solution to capture questionnaire data, including a standardized method to geographically reference each passenger's place of origin and travel destination, can help save the public health community valuable time in understanding the transmission dynamics and potentially lessen the spread of the disease.

GIS also plays a key role in supplementing traditional contact tracing. Geographically referencing contact information allows hospitals to perform location analysis to identify places in the community at higher risk of transmission and potential points of incidental infection (when the contact is unknown). GIS can help the public health community rapidly capture standardized and geocoded addresses for confirmed cases and case contacts in an effort called Community Contact Tracing. The effort provides essential support for attempts to slow the spread of disease throughout the community by breaking those disease transmission links among people and places.

Addressing this pandemic is part of Esri's common mission in bringing geographic science, GIS technology, and geographic thinking to every organization globally during these difficult and challenging times. ArcGIS applications such as Survey 123, StoryMaps, ArcGIS Pro, HUB, Tracker, Dashboard, and dozens of solution templates are helping people understand and tell their stories in real time. GIS application builders around the world have embraced the platform in the ongoing battle against COVID-19.

We have much more to do. Our way of living has changed, and we now need to think about what's next. What will the workplace look like in a physically distanced world? How can we make our communities more resilient (event planning, elections, etc.)? How do we balance these important decisions around economic and public health? GIS will play a crucial role in all these inherently location-based problems.

REOPENING THE WORKPLACE WITH INDOOR MAPPING

Leaders of businesses and organizations large and small faced multiple, complex challenges and decisions in their desire to reopen. In the United States, the White House released guidelines for reopening workplaces in phases, and those guidelines evolved through the summer as virus infections spiked in many states.

The ArcGIS® platform helps those tasked with providing a safe environment while adjusting to the evolving guidelines. Employers with large workforces and complex campuses showed particular interest in how to reopen facilities in compliance with federal, state, and local guidelines without jeopardizing employee health or wellness. With potential vaccines still in clinical trials through the summer, social distancing requirements, for example, required a new normal in the workplace. Some employers had to rethink floor plans, especially if their businesses used an open floor plan. Other businesses designated new walking routes to minimize worker interactions and avoid contaminated areas. Communicating with the workforce also is crucial, especially if an employee is exposed to someone who has tested positive for COVID-19. And all employees can expect to know what spaces are off limits and understand the sanitation schedules for their job sites. With employees spread across more buildings in some cases, employers face the challenges of closing off areas of high contact and ensuring regular and ongoing sanitizing to lower, if not eliminate, the risk of infection.

One federal guideline instrumental to the well-being of employees and the successful recovery of their community has proved daunting: "develop and implement policies and procedures for workforce contact tracing following employee COVID+ test." Tracing can be as simple as asking infected workers whom they came in contact with. Answering this question may be easy in a small business but might be a difficult task in a large company.

Tracking the movement of people outdoors relative to the virus has proven easier than contact tracing indoors. Even fewer technologies are capable of accurately representing contact between two employees within the walls and floors of a building. Esri, for one, has developed location tracking dashboards to support businesses as their employees return to work.

Protecting the workforce

Employers understand that the health of workers, their families, and surrounding community is vital to the global recovery and success of their businesses. As the virus spread globally, the workforce included employees in one of three categories:

- Those who got sick from the virus
- Those who did not know they were infected because they had minor or no symptoms (asymptomatic)
- Those who have not been infected

For this reason, COVID-19 prevention measures are essential at work so that employees can feel confident about their return until a vaccine is released.

Using a passive contact tracing solution provides employees with the peace of mind that interactions are monitored and that high-contact areas are routinely sanitized. Employers can see where traffic is greatest and can automate alerts to cleaning crews after a certain threshold has been reached. Maintenance crews can be monitored against cleaning goals to ensure that staffing levels are adequate to keep facilities clean.

With the ability to trace movement indoors and outdoors, public health and corporate security can meet their goals. This solution can scale to support thousands of employees across multiple sites to safely advance through each phase of recovery and enhance safety far into the future.

Understand proximity tracing

ArcGIS Indoors allows employers to deploy tracing apps for iOS and Android devices. With this capability, employers can provide tools for any device to aggregate workforce movements in real time and log contacts for historical analysis.

Performing spatial analysis on historical employee movement data enables employers to determine high traffic areas for sanitation and detect proximity among employees across space and time. If an infected employee gets past safeguards, the ability to quickly trace and identify their contacts can help quell the spread.

Dashboard showing intersecting location tracks.

MODELING THE CURVE

Esri Spatial Statistician Lynne Buie

Epidemiologists watched with growing concern early this year as the epidemic became a pandemic. They anticipated shortages of hospital beds, supplies, devices, and medical workers. Ahead of inquiries from concerned leaders, scientists began creating analytical models to quantify and predict the surge in COVID-19 cases and identify interventions to flatten the curve.

From that effort, several powerful models emerged as useful tools for hospitalization planning. Penn Medicine's Predictive Healthcare Team adapted the Susceptible, Infected, and Recovered (SIR) epidemiological model to create a new model it calls CHIME (COVID-19 Hospital Impact Model for Epidemics). The CDC created another new tool called COVID-19Surge, which uses a similar epidemiological model that takes into account more stages of the disease. The explosion of domain experts creating essential modeling tools has helped us better understand the potential impact of the pandemic. Incorporating geography into web and spreadsheet-based tools improves the modeling of complex COVID-19 phenomena. The GIS community is well placed to integrate spatial data into COVID-19 models; for example, metrics of social distancing and hospital capacity vary locally and have an important effect on the local outcome. Domain experts must visualize the model results geographically and communicate actionable information in intuitive applications and information products designed for hospital administrators, public health administrators,

emergency operations centers, and first responders. Esri integrated these two models into an ArcGIS Pro toolbox to help the community take advantage of the models and bring location into the workflow.

The two tools in the COVID-19 Modeling toolbox—CHIME Model and COVID-19Surge—estimate how many COVID-19 patients will need hospitalization, and of that number how many will need ICU beds and ventilators. The models can account for interventions such as social distancing and mandatory face mask policies or even simulate the impact of strengthening or relaxing these measures. Bringing these tools into a spatial analysis environment allows researchers to run the models for multiple hospital catchments or counties simultaneously, and adapt the model to specific disease patterns or policy decisions at each location. Using inputs such as total population, active cases, and currently hospitalized cases for each location, the tools produce spatial data showing anticipated hospitalizations, ICU hospitalizations, and ventilated patients for each day of the modeled period. Seeing the modeled curve helps hospital administrators plan ahead to meet forecast spikes in demand. The tool produces charts that visualize the modeled curve, estimating when hospitalizations, ICU admissions, and ventilator needs will reach their projected peaks. Modelers can configure these charts to show how single or multiple interventions help lower these peaks and inform policy makers on the impacts of proposed interventions.

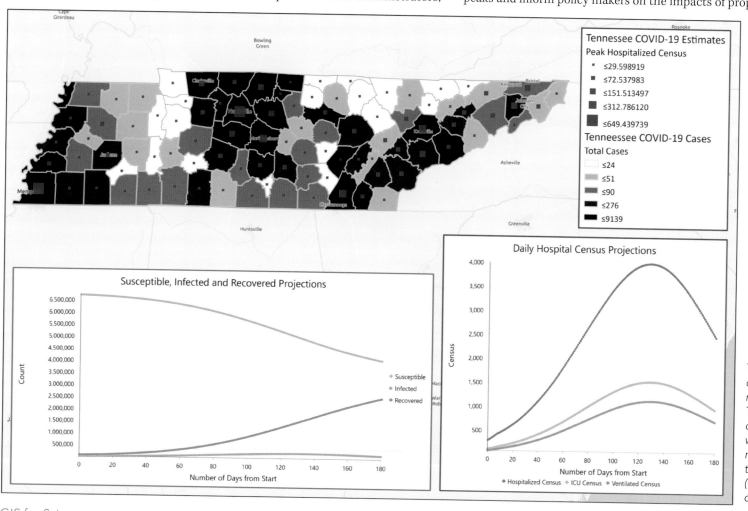

The ArcGIS implementation of the CHIME Model provides maps and visualizations that Tennessee state government officials could use to plan when and where they need to mobilize resources to lessen the impact of the surge. (Test data—not for policy or decisions.)

A single epidemiological model is not enough to model the complex phenomena of COVID-19: where one model suffers, another may excel. On the surface, the CHIME Model and COVID-19Surge appear to do the same thing—estimate hospitalizations—using similar types of epidemiological models. However, the critical figures that drive these tools are very different. The results of the CHIME Model tool are driven predominantly by the number of hospitalized COVID-19 patients and the amount of time it takes the disease to double in the population. The COVID-19Surge tool is instead driven by the number of cases and the number of new infections that have been observed per case. These differences may seem subtle, but the different inputs for each model, along with the slight differences in the type of compartmental epidemiological models used, can lead to big differences in the results of each tool. The data for one tool may be also easier to obtain—or less reliant on external factors such as testing—than the other. Therefore, it is important not to consider a single model in isolation. Models are only a simulation, and these simulations depend on the modeling techniques and data inputs. The more models we consider, the more certain we can be of our results.

Analysis and predictive modeling are most effective when policy makers receive actionable information on complex problems in a visually understandable format. To this end, the Capacity Analysis configurable app allows users to consume the information from the CHIME Model and COVID-19Surge tools in an interactive application in ArcGIS Online or ArcGIS Enterprise. The app focuses on comparisons across models: for example, an analyst may compare the results of the two tools for the same inputs or investigate different parameters using a single tool. The second approach allows analysts to investigate how proposed interventions, such as physical distancing, may impact modeled hospitalizations. By comparing the outcomes across different intervention scenarios, the CHIME Model tool, the COVID-19Surge tool, and the Capacity Analysis app together become an effective tool for decision-makers. ArcGIS Pro tools and configurable applications together help domain experts research and model the COVID-19 curve. The software also informs policy makers as they try to flatten the curve so that sick people don't overwhelm the capacity of our health care system.

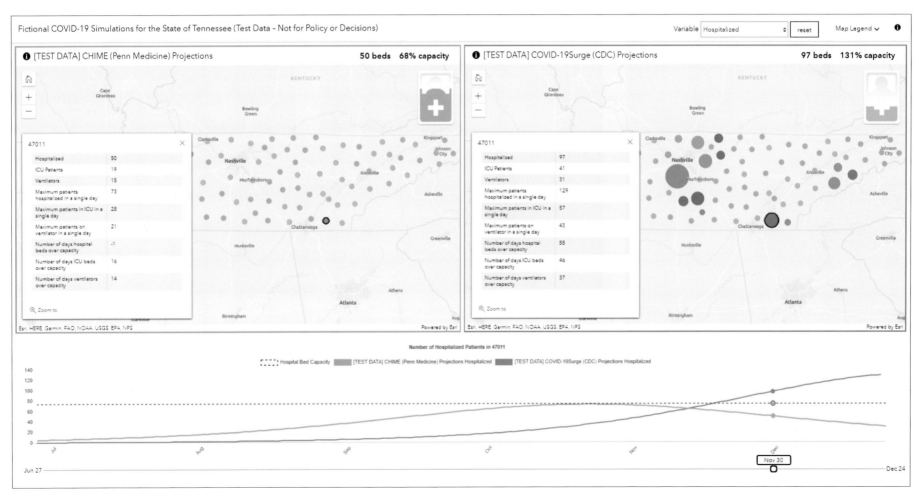

Analysts can put interactive results into the hands of nontechnical decision-makers using the Capacity Analysis configurable app, designed to work seamlessly with the results of the COVID-19 tools. (Test data—not for policy or decisions.)

THE RISE OF DASHBOARDS AND HUBS

In response to the pandemic, public health agencies and governments at all jurisdictional levels worldwide encountered intense demands for good data. As GIS professionals from across these myriad organizations scrambled to deploy useful and timely data-reporting tools, Esri globally began to put up resources and guidance about how best to use its technology, ArcGIS, an umbrella name that includes the full suite of Esri geospatial softare tools. These off-the-shelf tools uniquely enabled users to geographically organize, tabulate, visualize, and share COVID-19 data with the public.

First, Esri released the Coronavirus Response Solution, a package of two dashboard configurations—the Community Impact Dashboard and the Coronavirus Case Dashboard. This synchronization of data enabled people to easily deploy localized dashboards for their countries, states, and local provinces. The Community Impact Dashboard is designed to help public health agencies share basic COVID-19 testing and case metrics along with other key community information such as school closings and meal distribution sites. The Coronavirus Case Dashboard is designed to allow public health agencies to share more-detailed COVID-19 testing and case metrics with the public. Additionally, the Coronavirus Response solution includes mobile versions of both dashboards.

At about the same time, Esri put up its COVID-19 GIS Hub as a central repository of maps, datasets, applications, templates, and other GIS resources for creating and offering coronavirus-related resources (https://coronavirus-resources.esri.com).

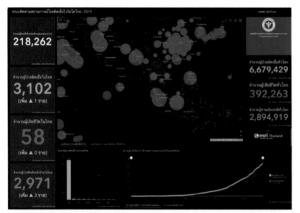

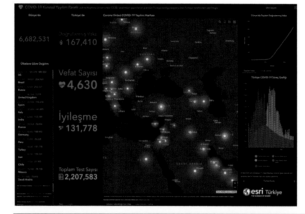

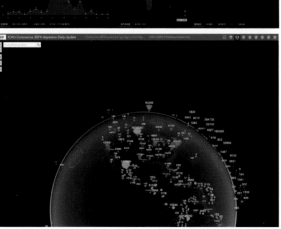

The COVID-19 GIS Hub remains an essential site for accessing current information and new applications for delivering virus data and is one of the largest GIS communities globally. Surrounding graphics illustrate the wide extent to which the GIS community deployed local dashboards around the planet.

A NEW ERA OF GIS

Esri Director of Products Clint Brown

In 2020, a new era of GIS exploded onto the global scene. Organizations everywhere banded together to form GIS communities aligned to focus on the world's great challenges. Just six months earlier, it would have been hard to imagine the power and reach of these ArcGIS communities; today their good work is everywhere to be seen.

By January, 2020, ArcGIS Online had already grown into a massive, shared-cloud GIS for the planet, containing more than 30 million spatially referenced information items covering literally every nation and corner of the world. What is most surprising—and what gives ArcGIS Online such a huge impact—is that more than 50% of that content is shared with other system users, forming an interconnecting web of content, projects, and efforts addressing every aspect of human activity. Public and private organizations have increasingly adopted and applied GIS as an integrating tool for understanding and action. This acceleration has long been a vision of leaders in the GIS world.

Historically, builders of GIS systems never felt like they had enough data, computing capacity, or bandwidth to fully realize the vision. Each GIS organization historically relied on data from other GIS users and organizations. Over time, computing and data-sharing networks continued to expand. Today, we see an environment where GIS organizations have created formal and informal alliances among themselves (based on shared geographies, shared topics, or both) to deliver their analyses and applications. As the cloud computing phenomenon exploded onto the information scene, more and more GIS organizations began to share their information and make it publicly accessible, so that more organizations could discover and link to this shared information and put it to work.

Everything happens somewhere. GIS is built on this premise; its organizing and interconnecting principle is *location*. GIS application builders use the rich location data that is a core part of every GIS as a foundation to integrate their own independent information layers. These sharing efforts have transformed GIS into a kind of magic tool for integrating content from multiple organizations. For decades now, many GIS organizations have collected and compiled these critical information layers, which can be easily combined and brought together. The result is a 180-degree shift: whereas the normal behavior was to silo data, today's progressive GIS organizations openly share, and the result is a quantum boost in the impact of their efforts.

A case in point emerged this year after the COVID-19 outbreak. A small engineering team at Johns Hopkins University (JHU) began to assemble their now widely recognized global COVID-19 dashboard. Early on, Dr. Lauren Gardner and the team made their dashboard (and the underlying data) public to support other scientists and medical professionals—adhering to the same ethics and open data principles that most GIS practioners would follow.

As the JHU team shared this dashboard, the news media picked up the web address, and the site began to immediately promote the JHU work. Gardner and her team followed some common data-sharing practices and ethics from the global geospatial community.

As of midsummer, 2020, the JHU COVID-19 dashboard had hundreds of billions of visits, the equivalent of dozens of views from every citizen on the planet. Just six months ago, it would have been hard to imagine the power and reach of the GIS tools and the good work provided by the JHU teams. The map became so ubiquitous that it is hard to imagine where the world would have been collectively if this application had not been built to access some sort of truth about the COVID-19 status worldwide. So why was this particular GIS application destined to become so accepted as the universally trusted tool used worldwide? A few thoughts come to mind. First, the small engineering team at Johns Hopkins that created the dashboard had the right philosophy:

- Make the site about sharing open information to provide the best available and up-to-date statistics.
- Respond intelligently. Pay attention to ongoing feedback about the information, and make verified corrections and updates as soon as possible. Build a cadence to maintain these updates and corrections.
- Instead of being defensive, acknowledge information errors and issues, and follow up.
- Learn how to be responsive. Continue to grow and evolve the solution strategically over time.
- Regardless of criticism and cyber-attacks, stay focused and remain open to feedback. Don't give up or give in.
- Continue to evolve and expand your offerings (e.g., the incredible work on US state and county maps and the dashboard's collection of county-based infographics [https://coronavirus.jhu.edu/us-map].
- Share your work and practice as a pattern so that others can emulate your results and set clear principles to follow.
- Maintain your commitment to support and sustain your information offerings.

At Esri, we see how this modern GIS experience ties back to all the efforts and investments that GIS organizations have made for many years. It's useful to realize that your solution incorporates a synthesis of content delivered as high-level information items from multiple sources—the whole is significantly greater than the sum of the parts. If you are a GIS practitioner, this spirit and ethos is in your blood.

GIS used to be almost entirely a back-office phenomenon, with highly trained professionals quietly laboring away using software and techniques only vaguely understood by the rest of us. The insights gained from their work benefited decision-makers within organizations but only occasionally reached larger audiences, and even then only as abstruse, static reports and posters. Suddenly, GIS is as much about communication as it is analysis. ArcGIS StoryMaps, ArcGIS Online, ArcGIS Dashboard, and Survey123 have turned GIS workers into communicators. GIS has burst out of the back office and has become accessible and actively used throughout organizations—and beyond.

Meanwhile, the COVID-19 pandemic has elevated the awareness regarding the role of GIS as a global tool for effective and sustainable community engagement. All this progress was made possible because of the best practices and ethics laid down by the earliest users of Esri's ArcInfo and ArcView (now ArcGIS and ArcGIS Online) communities.

GIS holds the promise of being a central component of a global network that can sense threats, map their extent, and help implement solutions. Climate change, environmental sustainability, and reduced biodiversity are three such global, existential threats. The great silver lining of the COVID-19 crisis is the possibility that we can apply the lessons we're learning even more broadly, with the ultimate and essential goal of achieving a sustainable and peaceful future.

WHY GEOGRAPHY STILL MATTERS

by Jared Diamond

Is geography increasingly irrelevant in the globalized world? Now that the internet, smartphones, and jet planes connect everyone to everyone else, has geography become unimportant, whether you live in Silicon Valley or the Central African Republic?

No, of course not. Just this year, millions of people—many for the first time—came to depend on maps and near real-time dashboards to stay abreast of a pandemic, not only to gain a sense of the global situation but to see how the new coronavirus affected their communities, even their neighborhoods. But geography is relevant in other ways, too. For example, you need money to afford a computer, smartphone, or airline ticket. Wealth is distributed unevenly around the world. Proportionately far more people in Silicon Valley than in the Central Africa Republic can afford a computer, smartphone, or airline ticket. First World countries on average are 32 times wealthier per capita than poor countries, and the richest countries, like Luxembourg, are 200 times wealthier than the Central African Republic.

But the effects of geography on national wealth hardly exhaust the importance of geography: they are just a first answer. What else can you think of that varies geographically, besides wealth? Climate change, of course. On average, the world is getting hotter, drier, and less productive agriculturally, and more at risk of fires. But some areas are getting cooler, wetter, and more productive, and less at risk of fires. (For example, compare California with Alabama within the United States or Australia with England in the larger world context.)

Resource problems also vary with geography—especially competition for seafood, timber, topsoil, and fresh water. For example, differences between China and Europe with respect to their peninsulas, islands, river configurations, and mountains shaped different political structures and technological innovation in China compared with Europe in the past, and they continue to do so today.

For now, let's look more closely at why geography is such a big reason for the differences in national wealth. If you have any doubts, you can easily see for yourself by doing this simple homework assignment: print out a map of Africa showing national boundaries. Look up online a set of numbers for the wealth of each country in Africa. You can use any of the usual measures of wealth that you prefer, many of them tabulated by the World Bank: average income per person, GDP per person, or GDP per person, corrected for purchasing power parity (i.e., differences in cost of living). Write those numbers for the national wealth of each country over the name of each country on your map. Compare the numbers at a glance.

Two conclusions will leap out at you. First, as far as the geography of wealth is concerned, Africa is a sandwich, with a thick core between two thin slices of bread. The core is the big tropical center of Africa, consisting of 38 countries. The two thin slices of bread are Africa's north temperate zone lying on the Mediterranean, consisting of five countries (Egypt, Libya, Tunisia, Algeria, and Morocco), and Africa's south temperate zone at the southern tip of Africa, consisting also of five countries (South Africa, Namibia, Botswana, Lesotho, and Swaziland). Compare the wealth of the 10 countries in those two temperate zones with the 38 countries in the tropical core. It will be obvious that most of the countries in those two temperate zones are wealthier than almost all the countries in the tropical core. (A nominal exception is Equatorial Guinea, which has an apparently high average income per person, because the president has an income of billions of dollars, while most other people in the small population have incomes of a few hundred dollars, so the average income looks high).

Evidently, living in the tropics comes with huge economic disadvantages compared with living in the temperate zone. One disadvantage of the tropics is low agricultural productivity, resulting from thin infertile soils and abundant insect pests and parasites that destroy crops. A second disadvantage is that chronic tropical diseases hurt the economy. People have shorter average lifespans, need more sick days, and stay home more often to care for their young, in part because families tend to compensate for the higher infant mortality rate by having more children. Finally, machinery is constantly breaking down in the heat. You can see this economic disadvantage of the tropics even in countries that span a wide range of latitudes from the tropics to the temperate zones, including Brazil and formerly the United States before air conditioning became widespread.

Another conclusion about geography leaps out at you from your map of Africa. Of Africa's 48 countries, 33 are along coastlines or on navigable rivers, but 15 are landlocked—either they have no coastline or cataracts block their rivers. Transport by boat is seven times cheaper than transport by air or by land. You will see that landlocked countries, regardless of location, are on the average about 40 percent poorer than countries with water transport.

If you still aren't convinced about this role of geography, and if you think that it represents a peculiarity of Africa, put the corresponding numbers for national wealth on a map of South America, which is simpler because there are only 12 countries to compare. You will see that the three richest countries of South America—Argentina, Chile, and Uruguay—are in the south temperate zone. Also just as in Africa, South America's poorest country—Bolivia—is its only landlocked country.

Of course, other factors besides geography affect national wealth. Those factors include corrupt institutions (although they too are ultimately influenced by geography and history), the so-called curse of natural resources (which paradoxically causes countries dependent on valuable natural resources to become poor rather than rich), the so-called reversal of fortune associated with colonial history (which has resulted in colonies that were formerly rich becoming predominantly poor today), and environmental degradation. Yes, these other factors are significant. But geography is one of the most important determinants of national wealth. Despite the internet, smartphones, and airline flights, geography still has a big effect on your pocketbook.

Do these maps mean that geography condemns tropical countries to a hopeless fate and that citizens of tropical countries should resign themselves to inevitably remaining poor forever? No, of course not. Just as a doctor's diagnosis can help you overcome illness through medical treatment or lifestyle change, geographers' diagnoses have also provided some tropical countries with recipes for achieving wealth. For example, if your country is in the tropics, don't base your economy on agricultural exports—leave them to temperate-zone countries like the Netherlands, the United States, Canada, and Argentina. If you are the president of a tropical country, invest heavily in public health. These lessons have enabled tropical countries such as Singapore, Malaysia, Thailand, Costa Rica, and Trinidad and Tobago in recent decades to climb out of poverty, and in some cases, achieve First World wealth.

So what if tropical countries are poor? That's unfortunate for them, but is it a concern for citizens of wealthier temperate-zone countries? Sixty years ago, the answer to that question would have been "no." Today, because of globalization, the answer is "yes" for at least three reasons.

One reason is the spread of tropical diseases from poor countries with low public health budgets to rich temperate countries via airline travel. Examples include the spread of AIDS, Ebola, Marburg, Dengue fever, cholera, and Chikungunya around the world on airliners. Climate change creates the added risk of establishing those tropical diseases in temperate countries.

Tropical disease spread isn't the only way in which globalization brings the issues of poorer tropical countries to wealthier temperate countries. A second way is that poverty creates support for international terrorism among desperately poor populations. A third way is that globalization has made immigration a permanent reality as citizens from poor countries seek better opportunities in wealthier countries. These citizens understand that their governments' promises to create wealth may take decades to materialize, if ever.

The power of these examples is that they put time and place and phenomena together to enhance our understanding. And within the computerized world of GIS, our comprehension and engagement with the world are further accelerated.

I trust by now that these examples—and there are many more—show that geography is as relevant as ever, if not more so, in our globalized world. It remains the foundation of our understanding through science, and through GIS for that science. Geography still matters—a lot.

Jared Diamond, professor of geography at UCLA, is the Pulitzer Prize-winning author of Guns, Germs, and Steel; *as well as* Collapse, The Third Chimpanzee, The World Until Yesterday, Upheaval, *and other best-selling books.*

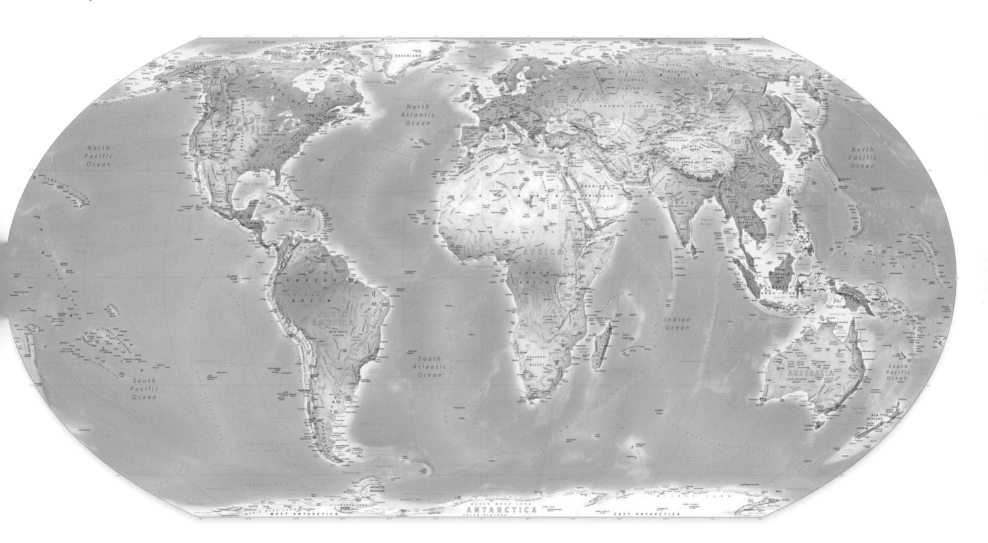

Equal Earth Physical Wall Map. The Equal Earth projection presents countries and continents at their true sizes relative to each other. Africa appears 14 times larger than Greenland, as it actually is. Visit equal-earth.com.

GIS for SCIENCE

APPLYING MAPPING AND SPATIAL ANALYTICS

Volume 2

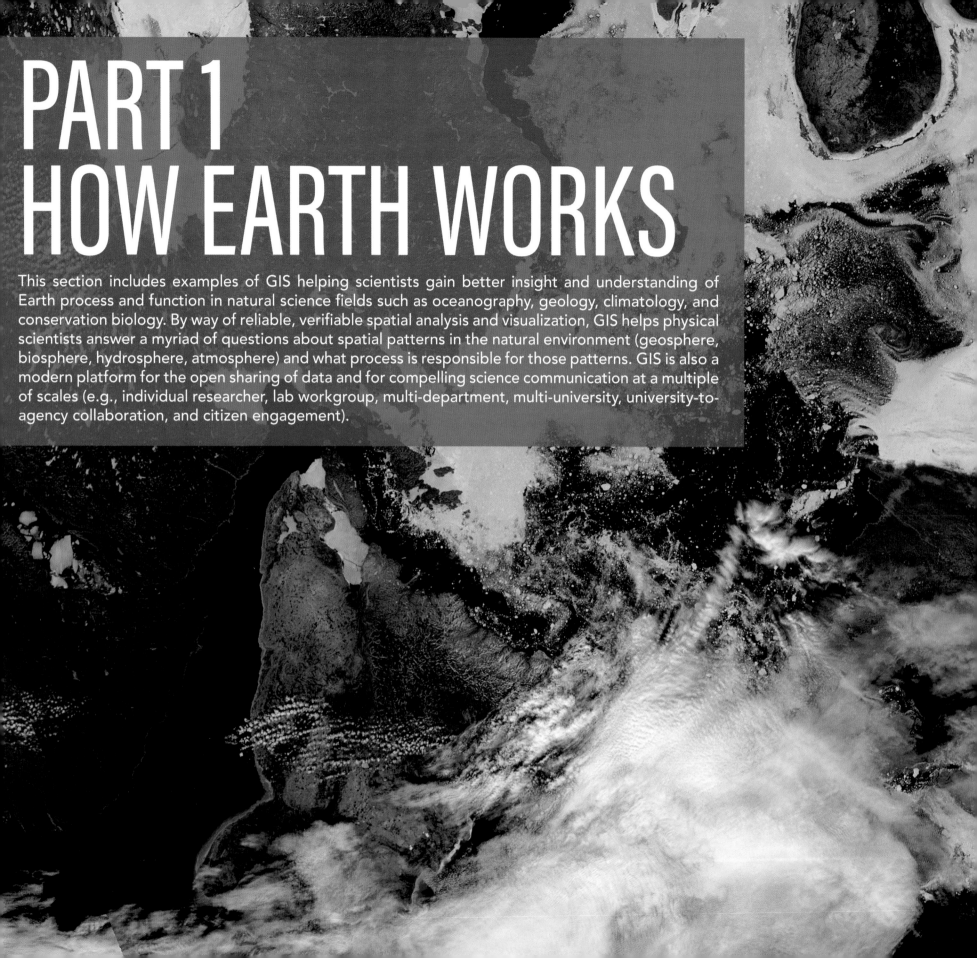

PART 1
HOW EARTH WORKS

This section includes examples of GIS helping scientists gain better insight and understanding of Earth process and function in natural science fields such as oceanography, geology, climatology, and conservation biology. By way of reliable, verifiable spatial analysis and visualization, GIS helps physical scientists answer a myriad of questions about spatial patterns in the natural environment (geosphere, biosphere, hydrosphere, atmosphere) and what process is responsible for those patterns. GIS is also a modern platform for the open sharing of data and for compelling science communication at a multiple of scales (e.g., individual researcher, lab workgroup, multi-department, multi-university, university-to-agency collaboration, and citizen engagement).

Mosaic of satellite pictures of Baffin Island from Aqua satellite.

THE GEOGRAPHY OF ISLANDS

Islands come in all shapes, sizes, and types, from tiny rocky outcrops to enormous continental landmasses. The true number of islands distributed in the planet's seas and oceans is still elusive. Recent efforts, bolstered by an abundance of detailed satellite imagery and the sophistication of geographic information systems (GIS), are bringing answers to those questions closer than ever.

By Roger Sayre, Madeline Martin, Jill Cress, **US Geological Survey**; Nick Holmes, **The Nature Conservancy**; Osgur McDermott Long, Lauren Weatherdon, **UNEP World Conservation Monitoring Center**; Dena Spatz, **Pacific Rim Conservation**; Keith VanGraafeiland, **Esri**; and David Will, **Island Conservation**

Tahanea is an atoll of the Tuamotu Archipelago in French Polynesia, a semi-autonomous state composed of 118 islands and atolls geographically dispersed over an expanse of more than 2,000 kilometers (1,200 miles) in the South Pacific Ocean.

ALL LANDS ARE ISLANDS

The word *island* is one of the more evocative words in any language. The word may bring to mind a tropical Caribbean paradise or suggest a remote polar mass of rock and ice. It may evoke a sense of place associated with home or a memory of a past visit across the waters. Some will think of island peoples and their cultures, while others may be drawn to thoughts of wonderful, rare, and sometimes endangered island animals and plants. When asked, "What is an island?," a typical response might be, "A small area of land surrounded by the ocean, with palm trees and sand." In reality, however, islands come in all shapes and sizes and types, from tiny islets no larger than rocky outcrops to enormous landmasses the size of the continents.

All landmasses on Earth, no matter how big, are surrounded by oceans and are therefore islands. That means we are all islanders. It is not a case of islanders versus mainlanders. We all live on islands, whether we see or feel that reality on a daily basis. For all of us, then, islands are our homes, so we must know them well and take care of them.

Surprisingly, given that islands are our collective homes, we are still seeking answers to basic questions such as, "How many islands are there on Earth, where are they located, and what are they like?" Despite many attempts to map and characterize islands across history, we still lack a definitive characterization. The true number of islands distributed in the planet's seas and oceans remains elusive. We still don't know exactly how much of the Earth's surface is made up of islands. However, thanks to the abundance of satellite imagery and the sophistication of geographic information systems (GIS), the answers to those questions are ever closer.

This chapter describes a recent partnership to map the islands of planet Earth. This characterization stemmed from a fruitful collaboration among government, private sector, academic, and nongovernmental organizations. The team used sophisticated geospatial analysis technologies to elaborate a new map of global islands at a 30-meter spatial resolution. What follows is a description of the work to merge two authoritative global island databases (GID) into one. This effort involved compiling island data from multiple sources, and reconciling and making the data available in the public domain as a free and open access resource. There is a solid realization that the planet's island systems—as the home to a great number of threatened and endangered species—have significant importance from a conservation perspective. High-quality and high-spatial resolution maps of the distributions of global islands are important for a variety of science applications, including analyses of species rarity and vulnerability, exotic species invasions, conservation priority, ecosystem value, sea level rise, and other investigations.

Matureivavao, the largest atoll within the Acteon Group, administratively part of the commune of the Gambier Islands.

Robben Island—approximately 6 kilometers off the coast of Cape Town, South Africa—a location best known for more than 400 years as a prison that held Nelson Mandela among other political prisoners. Today it is a UNESCO Cultural Heritage site.

Australia, the fifth-largest landmass on Earth.

GLOBAL ISLAND GEOGRAPHY IN ANTIQUITY

Early attempts at mapping global islands

Islands are shown on the earliest flat-Earth maps of antiquity, on the maps from the golden age of seafaring and exploration in the 15th and 16th centuries, and on the maps of the modern era. Imaginary islands often peppered early maps, a cartographic tradition stemming from what has been called *horror vacui* in Latin, an aversion to empty spaces on maps.

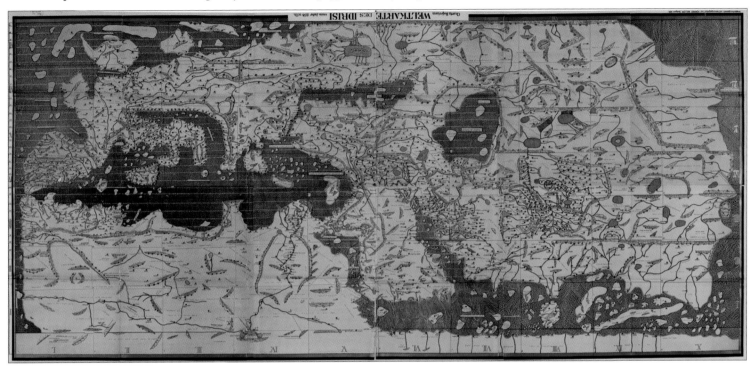

This map is an inverted Tabula Rogeriana flat-Earth depiction by Al-Idrisi, drawn in 1154. A technically competent cartographer, Al-Idrisi developed this map in what is known as a south-up orientation, possibly in an attempt to focus attention on the centrality and importance of Arabia. The Al-Idrisi map is often displayed in an inverted fashion, as here, to show the landmasses in the more common and familiar north-up orientation. While this map contains many depictions of real islands, recognizable by their shape, size, and location in spite of cartographic exaggerations, it also contains a number of imagined islands.

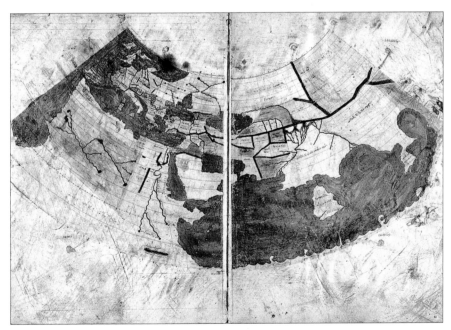

This map is the circa 1300 conic projection World Map by Ptolemy. The British Isles are depicted, as well as certain islands of the Mediterranean Sea, and what is likely Sri Lanka. This map was well respected and well used in its time.

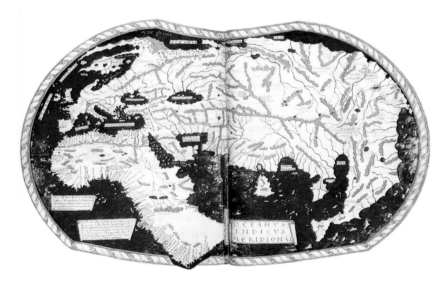

This Martellus World Map of 1489 drew significantly from Ptolemy's World Map but added many imaginary islands. Martellus' map appeared in his book, Insularium Illustratum (Illustrated Book of Islands), which contains detailed and rich maps of several Mediterranean islands.

Over time, and with increasing maps and knowledge from the accounts of the explorers, cartographers refined their depictions of the islands of the world. Meanwhile, geography rapidly evolved as a scientific discipline, with the emergence of sophisticated models of Earth as an irregular spheroid and numerous projections for representing its features on 2D maps. By 1800, the general locations, sizes, and shapes of the world's islands, the larger ones anyway, were well documented, as the next map shows.

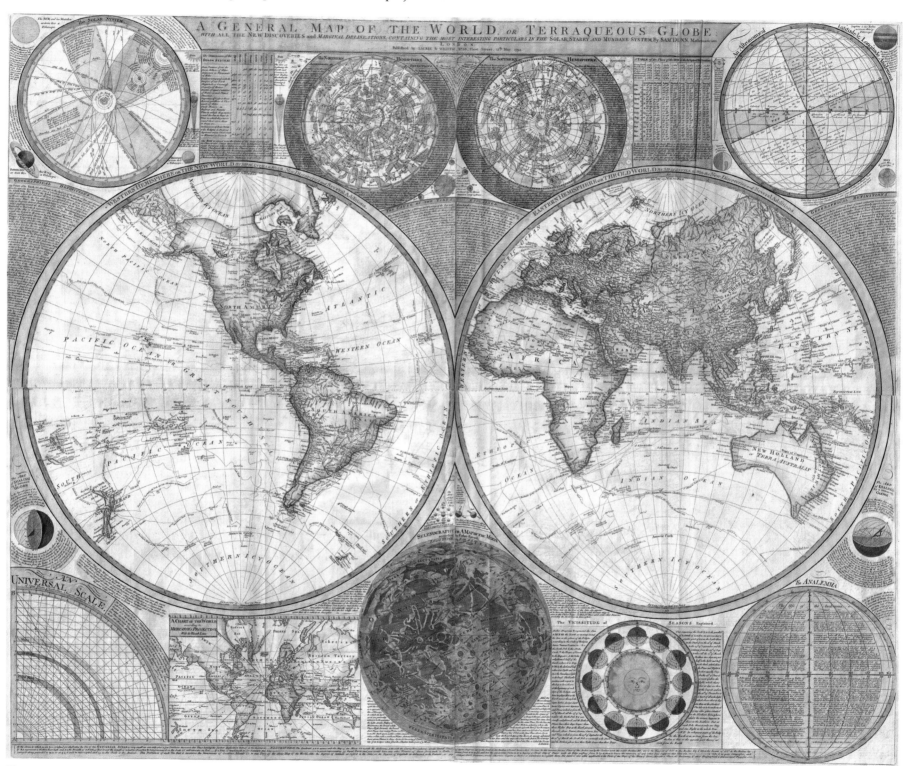

This map is an astonishingly detailed world map from 1794 by Samuel Dunn, with the comprehensive (and not at all mundane) title of A General Map of the World, or Terraqueous Globe with All the New Discoveries and Marginal Delineations, Containing the Most Interesting Particulars in the Solar, Starry and Mundane System.

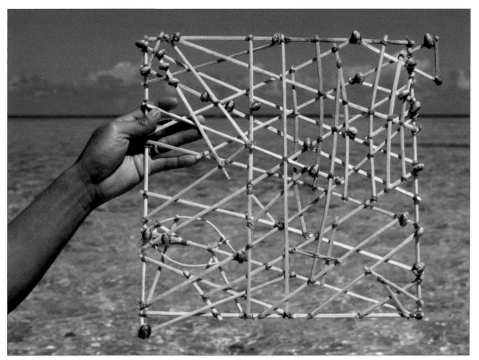

INITIAL MODERN COMPENDIA OF GLOBAL ISLANDS

With centuries' worth of accumulated cartographic representations of global islands available and not much in the way of terra incognita remaining to be discovered, modern geographers have turned their attention to detailed inventory and mapping of islands of increasingly smaller sizes. Prior to the ubiquity of GIS as a cartographic and analytical tool, geographers undertook global island inventories as compilations of existing information into standardized lists of islands, often sorted by size. Two pioneers of this kind of compendium development work were Dr. Arthur Dahl of the United Nations Environment Programme (UNEP) and Dr. Christian Depraetere of the French Research Institute for Development (IRD; formerly ORSTOM). They collaborated frequently to produce groundbreaking work on tabular databases and early GIS data layers on global islands. They developed rich attribute information on island names, physical geography, human geography, ecology, and special features. During the 1980s and 1990s, these resources were considered definitive compendia, and the UNEP Islands Directory[1] was available online in the early days of the web. It is still available at http://islands.unep.ch (note—the resource still exists online but has not been maintained since 2006).

GSHHS: The Global Self-Consistent Hierarchical High-Resolution Shorelines Map

In 1996, Paul Wessel and Walter Smith published the Global Self-Consistent Hierarchical High-Resolution Shorelines (GSHHS) database,[2] a game changer in the continuing effort to map global islands in a standardized manner. They used a digital coastlines dataset called the World Vector Shorelines (WVS) resource, digitized by the National Geospatial Agency (NGA) from nautical navigational charts. After considerable editing of the WVS to clean up aberrations in the vector linework and fill in missing coastline segments, they applied polygon topology to the shorelines to create a global islands GIS database. They used GSHHS data to delineate 180,500 islands, several orders of magnitude greater than the numbers of islands included in

the Dahl and Depraetere inventories (~1,000–2,000). For many years, this database has been considered both the original and definitive GIS data layer of global islands, with a large number of users.

IBPoW: The Island Biodiversity Program of Work

In 2006, the Convention on Biological Diversity at its eighth Conference of the Parties announced the first Island Biodiversity Program of Work (IBPoW) and associated Global Islands Partnership network. Dahl and Depraetere, in collaboration with UNEP's World Conservation Monitoring Center (WCMC), then developed a Global Island Database (GID v. 1.0) to be used as the IBPoW-endorsed reference layer. The GID is a merger of earlier island data produced by Dahl and Depraetere with the GSHHS.

Open Street Map® islands

During this time, the Open Street Map (OSM) resource became available. OSM is a remarkable crowdsoucing effort to provide detailed geographic information on a variety of features in an open source platform in the public domain. Local users can use this resource to modify existing information or add new information. Although users are most familiar with OSM street/transportation networks features, OSM also provides global shorelines and islands features. The OSM shoreline data were derived from a piecemeal interpretation of Landsat imagery conducted over several years beginning in 2006. A coastline extraction algorithm was used, and global coverage was ultimately achieved. Accuracy of the vector, called the Prototype Global Shoreline (PGS) [https://wiki.openstreetmap.org/wiki/Prototype_Global_Shoreline], is reported by OSM as variable and in need of improvement in many areas. The OSM user community is encouraged to improve the PGS, and guidance is provided for that crowdsourcing exercise.

PROGRESSIVE IMPROVEMENTS IN ACCURACY

The WCMC Global Island Database v. 1.0

As the creator and official steward for the GID v. 1.0, WCMC maintained and distributed the resource, which was composed of some 180,000 GSHHS-derived polygons with a minimum island size of 0.1 kilometer.[2] With increasing discovery and application of the new geodata resource, users identified certain inaccuracies. These were mostly related to a sometimes poor fit of the GSHHS polygon to the shoreline of an island when the GSHHS polygons were displayed on top of satellite imagery. This type of issue is demonstrated with the data drawn over the satellite imagery from Gaya Island, Malaysia.

The graphic shows the fit of the GID v. 1.0 (IBPoW) island shorelines (in yellow) to these islands. Importantly, these shorelines were derived from nautical charts, not satellite imagery. The inaccuracies in location, size, and shape of the GSHHS island shorelines suggested that an image-derived global islands map might represent a considerable improvement in accuracy.

The WCMC Global Island Database v. 2.0

Given the availability of the new OSM satellite-image-derived global islands resource, WCMC initiated the development of a new version of the GID, GID v. 2.0, replacing the polygons from the GSHHS with the new set of island polygons from the OSM product. This effort increased the number of islands represented from ~180,000 to ~400,000. Many of the new islands that resulted in v. 2.0 were smaller than the 0.1 kilometer^2 minimum island size of v. 1.0.

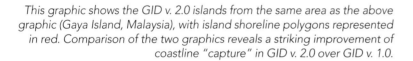

This graphic shows the GID v. 2.0 islands from the same area as the above graphic (Gaya Island, Malaysia), with island shoreline polygons represented in red. Comparison of the two graphics reveals a striking improvement of coastline "capture" in GID v. 2.0 over GID v. 1.0.

The USGS/Esri Global Islands Data Layer

In 2018, the US Geological Survey (USGS), in collaboration with Esri, produced a new, standardized, high-spatial resolution (30-meter) map of global islands interpreted from 2014 Landsat imagery.[3,4] The new data resource produced was in effect a "byproduct" of an effort to make a new global shoreline vector (GSV) for use in a global coastal ecosystem delineation and classification. The group did not set out to produce a definitive global islands map. But in applying polygon topology to the new GSV, the group recognized that a detailed new global islands map would be an outcome.

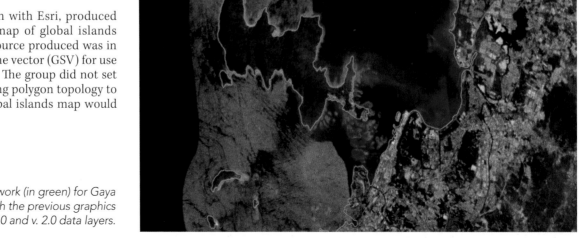

This graphic shows the new USGS/Esri islands line work (in green) for Gaya Island, Malaysia, to facilitate visual comparison with the previous graphics depicting the WCMC GID v. 1.0 and v. 2.0 data layers.

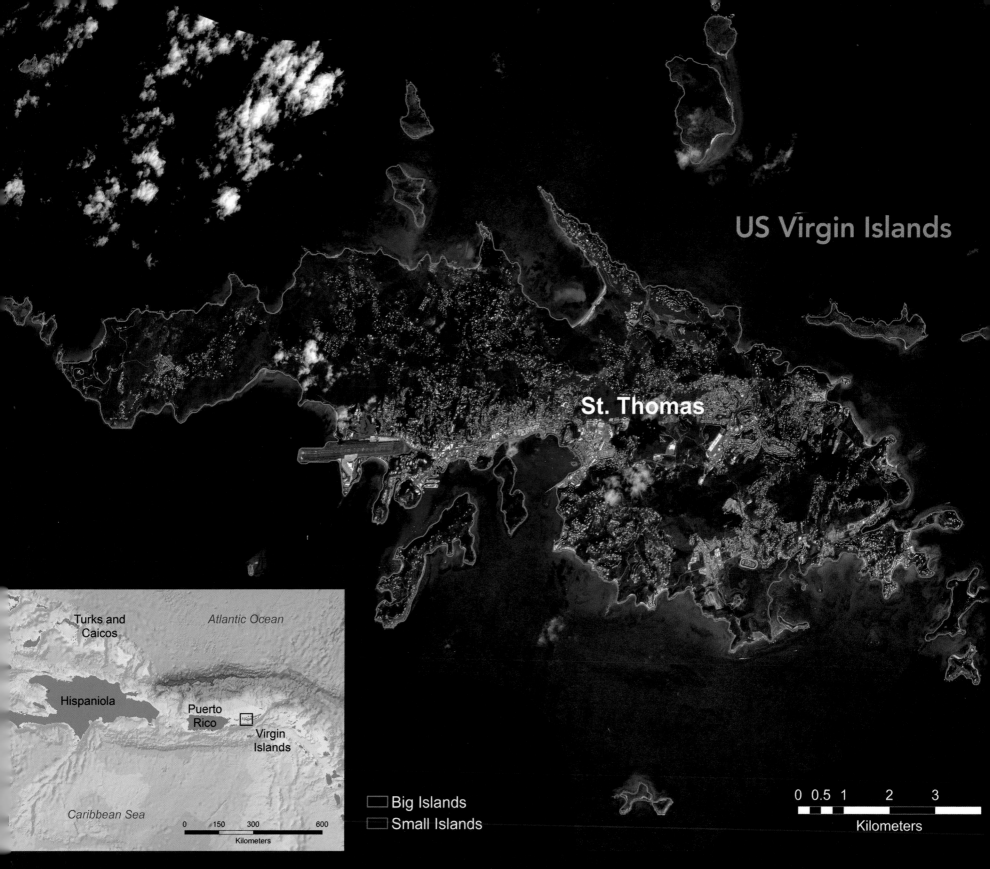

US Virgin Islands

St. Thomas

Turks and Caicos

Atlantic Ocean

Hispaniola

Puerto Rico

Virgin Islands

Caribbean Sea

0   150   300   600
Kilometers

☐ Big Islands
☐ Small Islands

0 0.5 1   2   3

Kilometers

The quality of the island polygon data is more easily evaluated when zoomed in to a fine spatial resolution, with the polygons displayed on top of satellite imagery. For example, in this graphic, the island of St. Thomas in the US Virgin Islands is shown with a green polygon outline surrounded by numerous smaller islets with red polygon outlines. A gestalt evaluation of the quality of the GIS data is provided from a visual inspection of the fit of the island polygon data to the shoreline of the island, as seen in the imagery.

The USGS/Esri Global Islands Data layer

The USGS/Esri global islands data layer was developed with a minimum mapping unit of 3,600 meters^2 (the size of four contiguous 30-by-30 meter Landsat pixels). The product, developed directly from semi-automated satellite-image interpretation, has few attributes. One important attribute is size class, with all islands identified as either Small Islands (< 1 km^2), Big Islands (> 1 km^2), or one of the five Continental Mainlands (North America, South America, Africa, Eurasia, and Australia). Names were added for all islands greater than 1 km^2 by a combined automated (intersection of polygons with the GeoNames geographic place-names data) and manual (analyst-based search for names using online mapping resources) approach. Full details of the methodology and results from the USGS/Esri global island data development effort are found in Sayre, et al.[3,4]

Basic characteristics of the USGS/Esri island polygons are found in the following table (reproduced from Sayre, et al.)[3]

Landmass type	Number of polygons	Area (km^2)	Length of coastline (km)
Continental mainlands	5	125,129,046	813,467
Big Islands (> 1 km^2)	21,818	9,938,964	1,304,762
Small Islands (≤ 1 km^2)	318,868	20,589	321,774

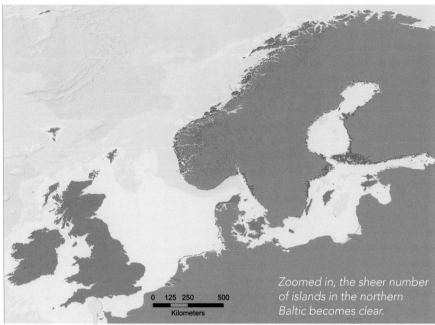

Zoomed in, the sheer number of islands in the northern Baltic becomes clear.

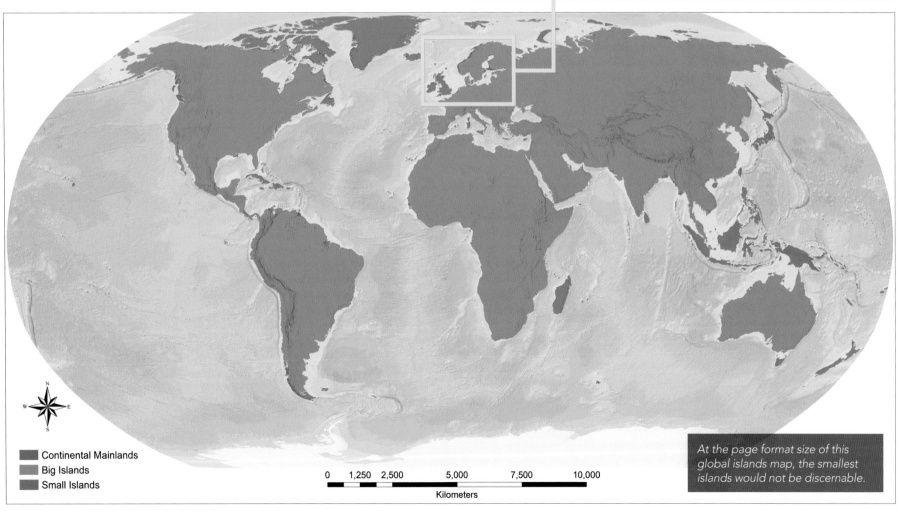

Continental Mainlands
Big Islands
Small Islands

At the page format size of this global islands map, the smallest islands would not be discernable.

Visualizing the USGS/Esri global islands data

To enable easy visualization and query of the USGS/Esri islands data by anyone with internet access, the USGS and Esri developed an online tool called the Global Islands Explorer (home page is pictured here).

Accessible at https://rmgsc.cr.usgs.gov/gie, this tool offers pan, zoom, and query functionality and allows the display of island polygons over a number of basemap backdrops, including satellite imagery and topographic maps. The island data are served as raster image services but are available for download in the tool in their original vector polygon format.

The global islands data have been placed into the public domain, accessible at: https://rmgsc.cr.usgs.gov/outgoing/ecosystems/Global/. The free availability of the data to anyone with a need or interest to use it is a testament to the open data philosophy that also resulted in the decision to make all Landsat data openly and freely available.

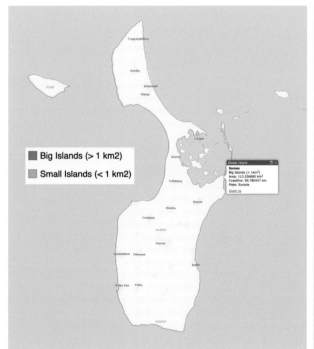
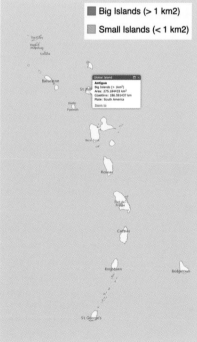
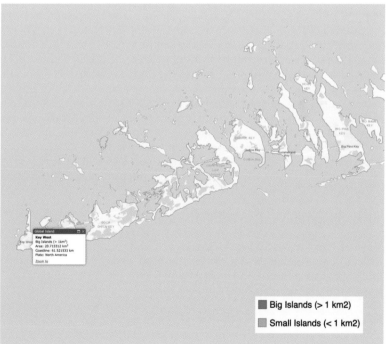

Three examples of islands and island chains as viewed in the browser-based Global Islands Explorer.

MERGING THE USGS/ESRI AND WCMC GLOBAL ISLANDS DATASETS

The arrival on the scene of the USGS/Esri product created an issue for seekers and users of global island data. First, how do two (USGS/Esri and WCMC v. 2.0) authoritative, spatially explicit, global islands resources available at similar spatial resolutions differ? And second, which one is more suitable for a particular application of interest? Wary of the burden this situation placed on users of global islands data, USGS/Esri and WCMC decided to reconcile the two resources to the extent possible into one. They agreed to merge the two resources to obtain an improved product and then place that resource squarely in the public domain for the greater good. This collaborative merging of resources is now complete. As is always the case when trying to reconcile datasets of differing origins, spatial resolutions, and intended uses, the merger was not perfectly straightforward and required use of sophisticated geospatial processing.

The USGS/Esri global islands as the reference foundation

The first and one of the most important decisions when contemplating the merger of the USGS/Esri and WCMC global island datasets was which one to use as a base layer for providing the foundational linework for the final combined database. Although the optimal approach would have been to examine corresponding polygon pairs for all islands and then select the best one to keep in the merged dataset, the enormity of that task, given the number of islands on the Earth (more than 300,000), precluded that analysis. We therefore needed to choose one data layer from the two global island products as the initial source of island polygons to work from.

The team visually evaluated both resources on the basis of accuracy, consistency, and level of detail in the linework when zoomed in to the spatial resolution at which the island polygons were originally interpreted (30 meters). After much globally comprehensive and rigorous visual inspection of the two resources, project leaders determined that the USGS/Esri resource would serve as the reference foundation and would be subsequently enriched using the WCMC data. That decision resulted in part due to the emerging realization that the USGS/Esri island polygons were more consistent globally, and because they had been produced in a documented and reproducible manner. Moreover, in most comparisons, the USGS data were more detailed in shoreline configuration. The WCMC linework (sourced from OSM) varied in consistency and detail from place to place, not entirely unexpected given the crowdsourced contribution for some of that linework. In some areas, the OSM-sourced linework lacked sufficient detail, with geometric shapes such as triangles and paralellograms representing island perimeters. For all these reasons, the USGS/Esri island polygons became the starting point for the merged linework product.

Initial preparatory edits along the continental mainlands coastline

The first step removed polygons from the WCMC data that conflicted (intersected) with any coastlines of the five continental mainlands in the USGS/Esri resource. All the USGS/Esri polygons, including the five large continental mainland polygons, were mutually exclusive from a spatial perspective, with no overlap of islands and continental mainlands. However, of the total initial number of WCMC island polygons (448,036), 42,787 polygons overlapped (intersected) the USGS/Esri continental mainland polygons and were therefore removed from the WCMC resource. Many of these islands were found to be located in interior freshwater lakes.

Use of the Near command to identify matching polygons in the data layers

The next step in the merge process identifed the subset of polygons in the WCMC data that matched (i.e., represented the same island) the corresponding polygons in the USGS/Esri data layer. Given that the USGS/Esri global island data had limited attribution (typically just name and size), the productiion team wanted to find matching polygons in the WCMC data to extract useful attribute information from the WCMC data and transfer it to the USGS/Esri polygons. We identified matched polygons as a pair, one from each data source, which each corresponded to the same island. This determination was made using the Near command in ArcGIS®. Near calculates distance and additional proximity information between the input features and the closest feature in another layer or feature class. We used the Near command to find the closest polygon in the WCMC dataset to the polygons in the USGS/Esri dataset within a specified search distance.

When the Near command returned a value of zero, the polygons in the two datasets overlapped, and an assumption was made that they matched. That assumption was borne out after a considerable number of initial visual comparisons revealed that in almost all cases the polygons were a matched pair. For the Near command, search radius was set to 300 meters based on the rationale that if a WCMC polygon was greater than 300 meters away from the USGS/Esri polygon, it may not represent a corresponding polygon for the same island. When the Near command returned a value of minus 1, it meant that no polygons were found in the search radius. For all matched polygons, the attribute information from the WCMC polygon was transferred to the USGS/Esri polygon.

The matched polygons

In all, 201,674 USGS/Esri islands directly matched to a WCMC island because their polygons overlapped. The country and name attribute information of the WCMC data was then joined to the corresponding USGS/Esri polygon. An additional 48,317 USGS/Esri island polygons were found to have a WCMC polygon in very close proximity (within the 300-meter search radius). Those nonoverlapping polygon pairs were also assumed to be a match, with the displacement attributed to differences related to projection dynamics, methodological differences, or errors in data creation. The Near command therefore successfully identified a total of 249,991 matched pairs from which WCMC attribute information could be extracted and transferred. A total of 90,690 USGS/Esri polygons did not have a WCMC polygon in the 300-meter search radius, so these polygons are assumed to be "missing" a WCMC counterpart, and thus lack additional attribution at this time.

Addition of WCMC polygons that were missing in the USGS/Esri data

At this point in the process, the merger of the two resources added considerable attribute information from the WCMC islands to the existing USGS/Esri islands—another powerful enrichment of the USGS/Esri islands data using the WCMC data related to island polygons in the WCMC data that did not exist in the USGS/Esri data. To find these WCMC islands that were "missing" in the USGS/Esri data, we ran the Near command in reverse, this time starting with a WCMC polygon and searching in a 300-meter radius for the nearest USGS/Esri polygon. WCMC polygons that did not have a match (Near analysis returned value of -1) were considered islands that potentially needed to be added to the USGS/Esri dataset. A total of 36,197 WCMC islands polygons had no match to the USGS/Esri data.

The team was only interested in adding those WCMC islands that did not exist in the USGS/Esri database if, in fact, they were real islands that had not been captured in the USGS/Esri image-based extraction. Because the WCMC metadata had a disclaimer warning of the existence of "fake" islands, it became necessary to verify that the polygons being merged into the USGS/Esri dataset actually represented real islands. We did not find a suitable automated method for testing the veracity of these WCMC polygons and determined that manual verification was the best and surest evaluation approach. We therefore visually inspected each of the 36,197 WCMC polygons over satellite imagery and marked them for inclusion in the USGS/Esri dataset if the analyst decided that the polygon represented an actual island.

While time and labor intensive, the process effectively identified anomalies and errors in this set of WCMC polygons. Sometimes, deciding whether the polygon was real was straightforward in that land, rock, sand (emerged), or vegetation was discernible. Other times, the decision was quite difficult, as swirling waters and whitecaps indicated the probable existence of rocks just below the surface of the water. The decision was made more difficult when the polygon was not overtop a land feature, but nearby (displaced). When the displacement was not considerable, and the size and shape of the polygon approximated the size and shape of the land feature seen in the imagery, the polygon was determined to be "real" (in other words, it represented an island) and was subsequently added to the USGS/Esri resource.

The team added any WCMC polygons with Antarctica as a country attribute to the USGS/Esri resource. This resource does not otherwise include Antarctic islands because of the lack of imagery for that region during the satellite image interpretation step. However, in the merged product, an Antarctic "mainland" polygon is still lacking.

Most human-engineered structures, such as seawalls, were excluded from the USGS/Esri dataset; however, certain artificial islands constructed to resemble islands, such as the Palm Islands off the coast of Dubai, UAE, were included in the dataset. Islands that were created to be islands were included, while lands built to support coastal infrastructure were not.

Very small islands (less than 3,600 square meters)

Many islands from the WCMC resource that were added to the USGS/Esri islands dataset were very small. Their small size explains why these islands were missing from the USGS/Esri resource in the first place. The minimum mapping unit used in the initial extraction of USGS/Esri island polygons from satellite imagery was 3,600 square meters, which is the area of four contiguous 30-meter-by-30-meter Landsat pixels. These very small islands from the WCMC data often delineate rocky outcrops or islets in the surf surrounding larger islands. Some many-to-one and one-to-many errors were noted at this very fine level of resolution wherein some single polygons actually represented a cluster of islets, or a cluster of island polygons represented a single island feature. Moreover, the detail in the linework of many of these small WCMC islands was generally less than the detail of the USGS/Esri linework.

For these reasons, researchers had less confidence in the detail and accuracy of the WCMC-sourced polygons than in the original USGS/Esri-sourced polygons, and users were encouraged to verify the accuracy of any polygons of interest added from the WCMC resource to the USGS/Esri resource. In general, users who want to use the data at a localized scale should verify the accuracy of polygons in this class and may want to make edits or adjustments to the linework as needed.

Characteristics of the merged database

During the transfer of matched-pair attribute information from the WCMC resource to the USGS/Esri reference, 249,990 islands were updated. Subsequently, an additional 28,727 islands from the missing island analysis were added from the WCMC resource to the USGS/Esri reference, and 12 USGS/Esri islands were removed. The new merged data layer now contains five Continental Mainlands, 22,471 Big Islands (larger than 1 km^2), and 346,925 Small Islands (less than 1 km^2). The total number of islands in the merged resource at the time of this publication is therefore 369,401, although this number may change slightly based on future refinements to the resource.

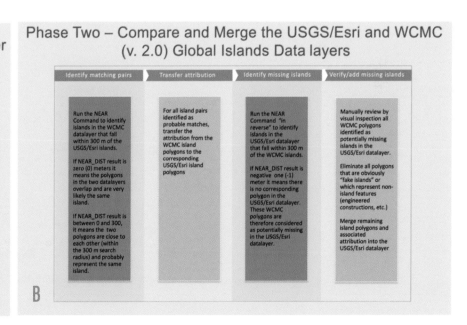

These workflows summarize the detailed methodological descriptions provided earlier. The diagrams depict (A) the development of the USGS/Esri Global Islands Data Layer, and (B) the subsequent harvest of additional information from the WCMC global islands product into the USGS/Esri reference layer. The new data are available in the ArcGIS Living Atlas and also in the public domain at https://rmgsc.cr.usgs.gov/outgoing/ecosystems/Global. A digital object identifier (DOI) has also been assigned to the data: https://doi.org/10.5066/P9C6XKL0.

ISLANDS WITH THREATENED BIODIVERSITY—A CRITICAL CASE STUDY

Many practical analytical applications require good, useful data on the location of islands, as well as their numbers, shapes, sizes, and features. These applications range from analysis of sea-level change to assessments of economic and noneconomic value of island ecosystem goods and services to identification of suitable areas for watercraft navigation. One extremely important application requiring good island distribution information is the study of island biodiversity. We now turn our attention to a critical case study—threatened island biodiversity—and discuss how GIS analysis supports the understanding of the conservation importance and status of island biodiversity.

Island biodiversity and islands as epicenters for species extinctions

Islands total only a small fraction of our planet's land area,[5] yet host extraordinary concentrations of unique species and are home to a disproportionately higher amount of the world's biodiversity than continents.[6] For example, of the more than 10,000 bird species described in the literature, 17% occur only on islands.[7] Many island species are found only on one island or island group and are thus considered endemic to that location. Madagascar, one of the largest of the oceanic islands, is home to as many as 15,000 native species of vascular plants, with 85% of them endemic. Endemism on remote oceanic islands results from the evolutionary adaptation of

founding populations of ancestral species that arrived from continents—via flight, oceanic flotsam, or other natural circumstance.[6] In the Hawaiian Archipelago, the establishment of a cardueline finch from the continent gave rise to nearly half of all the Hawaiian landbirds known today as Hawaiian honeycreepers, with more than 50 species, each with a different bill morphology and tongue shape to exploit diverse food sources—seeds, fruit, insects, and nectar. These birds provide an astonishing narrative of evolution on islands.[8]

Sadly, islands have been and continue to be epicenters for extinctions. Of the 275 total vertebrate extinctions worldwide since the 1500s, 54% of amphibians, 81% of reptiles, 95% of birds, and 54% of mammals were island species.[5] Extinctions are not a thing of the distant past. In 2012, the Christmas Island Pipistrelle (*Pipistrellus murrayi*) was declared extinct. This small bat underwent a rapid decline from 1994–2005, but conservation action was too late, and the last Pipistrelle call was detected in 2009.[9] Islands provide critical refuges for species at risk of extinction today. A study of 2,919 terrestrial vertebrate species classified as Critically Endangered (CR) or Endangered (EN) by the International Union for Conservation of Nature (IUCN) Red List of Threatened Species, a global scorecard for species conservation, found 1,189 (41%) breed on islands, highlighting the disproportionate number of threatened island species compared to continents when considering land area.[10]

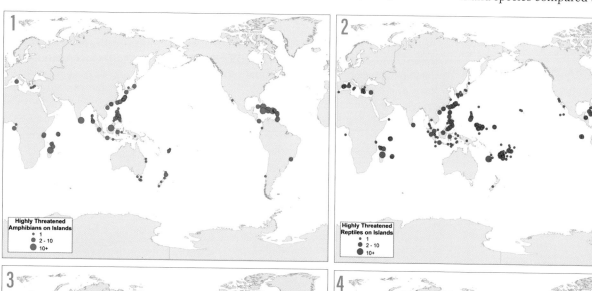

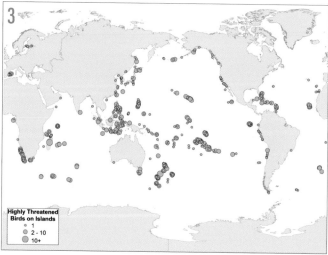

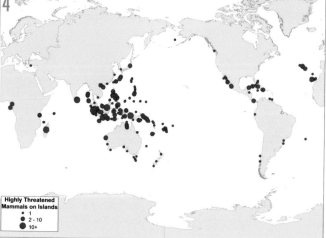

Relative numbers of threatened species on global islands for four taxonomic groups: (1) amphibians, (2) reptiles, (3) birds, and (4) mammals, from Spatz et al. 2017.[10]

Birds in danger

The Floreana mockingbird (*Mimus trifasciatus*) has been extirpated from the majority of its original home range, the island of Floreana in the Galapagos. It is likely that Darwin's observations of this bird species, also called the Charles mockingbird, strongly influenced his views on evolution and the origins of species. The mockingbird now exists in very low numbers on a few nearby rocky outcrops. Predation by dogs and cats eliminated the bird from Floreana. Sadly, this situation is increasingly common. The numbers of the struggling Polynesian ground dove (*Pampusana erythroptera*), a French Polynesian single island endemic, have been reduced to only about 100 individuals because of feral cats and invasive rats.

The Floreana mockingbird (Mimus trifasciatus) is a critically endangered bird now found on only a few offshore islets in the Galapagos Islands, having been eliminated from its namesake island by invasive rodents and feral cats.

The critically endangered Polynesian ground dove is endemic to the Tuamotus Islands of French Polynesia. It is found on just a few atolls and is now extirpated from several islands where it formerly occurred. Predation by feral cats and invasive rats has reduced its numbers to approximately 100 individuals.

Invasive species and promising eradication efforts

Invasive species have been a major driver of species losses, implicated in 86% of island extinctions.[11] Island species often evolved in the absence of native predators and herbivores, leading to high vulnerability upon first contact with humans, and ultimately the extinction of many species. On Midway Atoll in the Pacific, the smallest invasive mammal, the house mouse

Predator damage to an endangered blue-footed booby egg.

(*Mus musculus*), recently adapted to prey upon and kill the largest of seabirds, the albatrosses. Incubating adults, ecologically naïve to this threat, sit tight on their nests and are attacked relentlessly by mice, often with fatal consequences.[12]

Nonetheless, islands offer hope that we can prevent extinctions and protect biodiversity. The development of techniques in New Zealand to control or eradicate invasive mammals from islands has led to remarkable conservation success stories, and these techniques are now used around the world.[13] On Palmyra Atoll in the South Pacific, eradicating invasive Pacific rats removed a nonnative herbivore and seed predator, allowing a 5,000% increase in native seedling growth,[14] led to the extirpation of the Asian tiger mosquito (*Aedes albopictus*) by removing the primary host,[15] and created safe habitat that could host translocated populations of rare birds elsewhere in the Pacific. Islands within islands have also been created—for example, the development of predator-exclusion fences that keep invasive mammals out of important habitats on larger inhabited islands. The establishment of a predator-exclusion fence in Nihoku, on the island of Kauai, Hawaii, is supporting conservation recovery efforts for endangered Hawaiian shorebirds. The application of social attraction techniques, such as broadcasting albatross calls from a sound system and deploying albatross decoys in breeding displays, will further augment population recovery in these invasive-free sites.[16]

A predator-exclusion fence in Nihoku, Hawaii, constructed to safeguard important bird foraging and breeding habitat.[12]

APPLYING GIS AND DATA SCIENCE TO ENHANCE BIODIVERSITY CONSERVATION ON ISLANDS

Spatial assessments of islands and island biodiversity are essential to prioritize conservation planning and track conservation interventions. To realize these goals first requires a globally consistent spatial dataset that uniquely identifies each of the islands of the world. The World Conservation Monitoring Center (WCMC) Global Island database (GID) v 1.0 and v 2.0, and now the merged USGS/Esri and WCMC datasets, have provided a foundation to undertake these broad-scale conservation science investigations, two of which include the development of the Threatened Island Biodiversity Database and the database of Islands and Invasive Species Eradications.

The Threatened Island Biodiversity Database

The Threatened Island Biodiversity (TIB) Database (http://tib.islandconservation.org), created in partnership with Island Conservation, University of California, Santa Cruz–Conservation Action Laboratory, BirdLife International, and the IUCN Invasive Species Specialist Group, is the most comprehensive global review of island species listed as threatened on the IUCN Red List and at risk from invasive vertebrates,[17] and is considered the "gold standard" for filling biodiversity data gaps.[18]

The TIB documents the current and historical distributions of highly threatened animals, representing 41% of all critically endangered (CE) and endangered (EN) birds, mammals, reptiles, and amphibians on the planet.[10] These highly threatened animals were breeding on just 1,288 islands, representing only 0.3% of the ~400,000 islands worldwide, and with 70% of species restricted to a single island, representing hotspots for biodiversity conservation efforts.

The TIB was achieved by extensive literature review and consultation with more than 500 experts. The dataset was collated by first assessing all vertebrate taxa classified as CR or EN from the IUCN Redlist for breeding populations on only islands, on both islands and continents, or only on continents. For each island species, every unique island that hosted a breeding population was identified, documenting the present and historical breeding status for each population on each island, and linked to the WCMC GID. For each of these breeding islands, the presence or absence of invasive vertebrate species—primarily invasive mammals known to be highly damaging—were collated. Combined, the data allow conservation planners to identify and prioritize feasible conservation actions, such as prevention, control, and eradication of invasive species, to save island species from extinction.

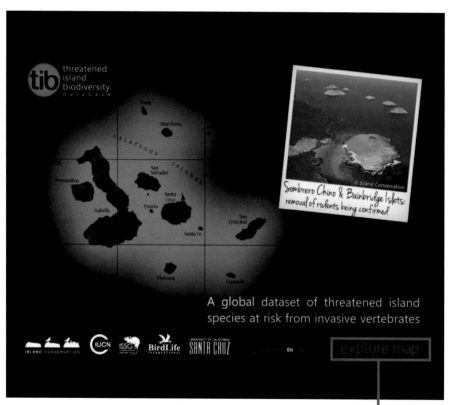

The Threatened Island Biodiversity Database is the most comprehensive global review of island species listed as threatened on the IUCN Red List and at risk from invasive vertebrates.

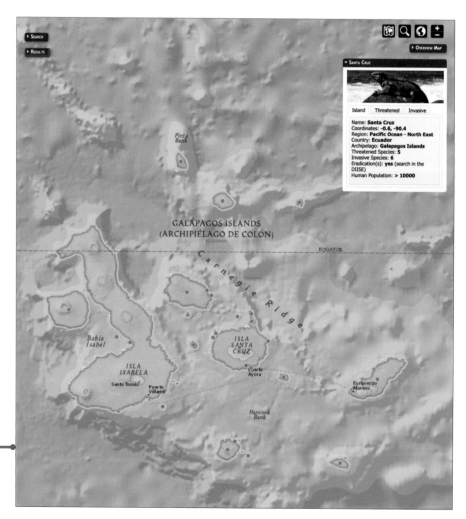

The Database of Island Invasive Species Eradications (DIISE)

The Database of Island Invasive Species Eradications (DIISE, openly available online at http://diise.islandconservation.org), created in partnership with the UCSC-CCAL, University of Auckland, IUCN ISSG, Landcare Research, and IC, compiles the methods and outcomes of invasive vertebrate eradication projects on islands.[19] To date, the database contains data on 1,400 efforts on 940 islands, of which 88% successfully met planned objectives. The database provides unique insight by tracking the global application and success rate of a conservation tool and provides important utility to conservation practitioners faced with managing the threats of invasive species on islands.

Creation of the DIISE is primarily based on systematic review of published and gray literature, and expert correspondence. Each eradication event described is linked to an island on the WCMC GID with a unique island code. Eradications on different islands were recorded as separate events, regardless of whether it was in the same archipelago or treated concurrently. Eradications of different species on the same island are treated as different events. Key parameters in the database include method, target species, outcome, and data quality.[20]

With help from the Center for Integrated Spatial Research at UCSC, the TIB database and the DIISE were published in 2012 and 2013 as publicly available web applications allowing users to identify islands from a series of parameters. Since their initial release, each database has undergone significant data updates. For earlier versions of the TIB and DIISE, spatial inaccuracies, omissions, and false islands in the GID v 1.0 required careful review of islands. Spatial representation of islands relied primarily on island centroids rather than polygons. Core island attributes, such as island area and degree of human habitation, were based on literature and expert review. With the development of the GID v 2.0, an extensive manual review process to "cross-walk" islands between the new and old datasets was undertaken and then validated using publicly available satellite imagery to correct island polygon size and location, resulting in more than 3,000 individual island polygons with rich attribution and high spatial resolution.

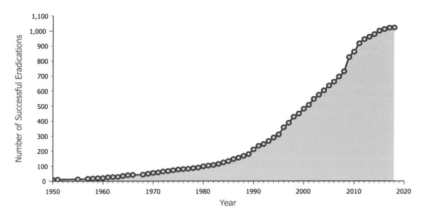

Cumulative number of successful invasive mammal eradication projects by year since 1950. Data are restricted to whole island events, where data quality is scored as good or satisfactory only, and excludes domestic animals and reinvasion events.

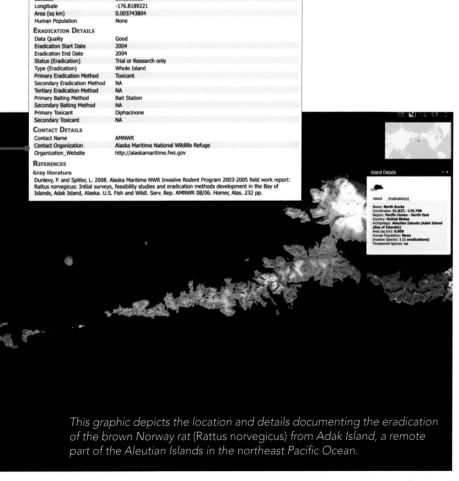

This graphic depicts the location and details documenting the eradication of the brown Norway rat (Rattus norvegicus) from Adak Island, a remote part of the Aleutian Islands in the northeast Pacific Ocean.

Spatial data: An asset for conserving threatened island biodiversity

Island habitats are unique and inseparably linked to traditional island cultural lifeways. There is a real urgency to protect habitats, many of which are among some of the last true wild places on Earth. The USGS/Esri/WCMC GID, the TIB, and the DIISE all provide insights about the biogeography of threatened island species, the success rate of island conservation efforts, and the prioritization of actions undertaken to prevent future extinctions. These databases have allowed for the identification of some 30,000 islands, a small but very important subset of the 369,401 global islands, that are known to harbor endemic and threatened biodiversity and merit increased conservation attention. Combined, these datasets and subsequent analyses have been cited in more than 70 peer-reviewed publications and have been integrated into national, regional, and global conservation funding and policy decision-making. Examples include identifying where globally threatened seabirds are at risk from invasive species and sea level rise,[21] where eradicating invasive mammals will benefit highly threatened vertebrates,[22] and what conservation outcomes have occurred following invasive mammal eradications.[23] Data on the number of eradications of invasive mammals on islands over time were also used within the Biodiversity Indicator Partnership, contributing to measuring progress toward Aichi Target 9 (tackling invasive species) within the Convention on Biological Diversity.

Global indicators

An indicator can be defined as a "measure based on verifiable data that conveys information about more than just itself." The United Nations Environment Program World Conservation Monitoring Centre (UNEP-WCMC) hosts the Secretariat to the Biodiversity Indicators Partnership (https://www.unep-wcmc.org/resources-and-data/biodiversity-indicators-partnership-global), an interdisciplinary global initiative to promote the development and delivery of biodiversity indicators. The Biodiversity Indicators Partnership exists to support the development of and reporting on indicators from a variety of intergovernmental protocols and agreements related to biodiversity and sustainable development. These protocols include the Convention on Biological Diversity (CBD; www.cbd.int/sp/targets/), the Intergovernmental Science-Policy Platform on Biodiversity and Ecosystem Services (IPBES; https://ipbes.net), the Sustainable Development Goals (SDGs; https://sustainabledevelopment.un.org/sdgs), and other conventions. Indicators have also been found to be useful for regional, national, and subnational reporting related to sustainable development and biodiversity conservation. Indicators are essential metrics for monitoring and reporting progress toward the achievement of national targets and are important in facilitating adaptive management.

An accurate global islands dataset will be important for the development of robust global indicators. Reliable and verifiable data is fundamental for the creation and maintenance of successful indicators. A high-resolution global islands dataset could be used to improve the accuracy of reporting units where islands intersect with terrestrial and marine environments, as for example, Aichi Biodiversity Target 11: "*By 2020, conserve at least 17 per cent of terrestrial and inland water, and 10 per cent of coastal and marine areas....*". A consistent and accurate islands dataset could help to standardize the reporting geographies necessary for the global policy instruments such as the CBD and the SDGs.

Data from the DIISE reporting the number of invasive vertebrate eradications on islands—underpinned by a global dataset of islands—have been used to measure progress toward Aichi Target 9: *By 2020, invasive alien species and pathways are identified and prioritized, priority species are controlled or eradicated and measures are in place to manage pathways to prevent their introduction and establishment* (Convention on Biological Diversity, 2011). This indicator is currently being used by the Biodiversity Indicator Partnership.

Conclusion

The mapping of islands from antiquity until today is an evolutionary tale of increasing geographic knowledge coupled with technological sophistication. Today, most of Earth's islands are known and have been mapped, but authoritative and high spatial resolution data on island distributions have been relatively lacking until recently. A merger of the USGS/Esri global islands data and the WCMC v. 2.0 global islands database has produced a new, detailed, and globally comprehensive islands geodatabase with more than 400,000 islands included. Users can easily explore these islands using a web-based visualization and query tool called Global Islands Explorer, and they can find freely available island vector polygon data in the public domain. An important subset (~30,000 islands) of the global islands harbor endemic and threatened biodiversity and merit conservation attention. The Threatened Island Biodiversity database and the database on Island Invasive Species Eradications are two GIS-based resources supporting the global island biodiversity conservation effort. Satellite imagery and geospatial technologies have enabled and facilitated the development and analysis of global islands data in general, and threatened island biodiversity in particular.

On this particular atoll of the Palmyra Islands, an effort to eradicate invasive species has met with considerable success. Two manta rays are seen swimming near the island.

NOTES

1. Dahl, A. 1991. *Island Directory*. UNEP Regional Seas Directories and Bibliographies Number 35. Nairobi, UN Environment Program. 573 pages.

2. Wessel, P., and W. Smith. 1996. "A Global, Self-Consistent, Hierarchical, High-Resolution Shoreline Database." *Journal of Geophysical Research*. 101(B4):8741–8743. doi:10.1029/96JB00104.

3. Sayre, R., S. Noble, S. Hamann, R. Smith, D. Wright, S. Breyer, K. Butler, K. Van Graafeiland, C. Frye, D. Karagulle, D. Hopkins, D. Stephens, K. Kelly, Z. Basher, D. Burton, J. Cress, K. Atkins, D. Van Sistine, B. Friesen, R. Allee, T. Allen, P. Aniello, I. Asaad, M. Costello, K. Goodin, P. Harris, M. Kavanaugh, H. Lillis, E. Manca, F. Muller-Karger, B. Nyberg, R. Parsons, J. Saarinen, J. Steiner, and A. Reed. 2019. "A New 30 Meter Resolution Global Shoreline Vector and Associated Global Islands Database for the Development of Standardized Ecological Coastal Units." *Journal of Operational Oceanography*, 12:sup2, S47-S56, DOI: 10.1080/1755876X.2018.1529714.

4. Sayre, R., J. Dangermond, D. Wright, S. Breyer, K. Butler, K. Van Graafeiland, C. Frye, D. Karagulle, S. Kopp, S. Noble, J. Cress, D. Burton, M. Martin, and J. Steiner. 2019. *A New Map of Global Islands*. Washington, DC: American Association of Geographers. 24 pages.

5. Tershy, B., K. Shen, K. Newton, N. Holmes, and D. Croll. 2015. "The Importance of Islands for the Protection of Biological and Linguistic Diversity." *Bioscience*, 65:592–597.

6. Whittaker, R., and J. Fernández-Palacios. 2007. *Island Biogeography: Ecology, Evolution and Conservation*. Second Edition, New York: Oxford University Press.

7. Newton, I. 2003. *The Speciation and Biogeography of Birds*. London: Academic Press.

8. Pratt, T. 2009. "Origins and Evolution." In Pratt, T., C. Atkinson, P. Banko, J. Jacobi, and B. Woodworth (editors), *Conservation Biology of Hawaiian Forest Birds: Implications for Island Avifauna*. New Haven: Yale University Press. p. 3–24.

9. Martin, T., S. Nally, A. Burbridge, S. Arnall, S. Garnett, M. Hayward, L. Lumsden, P. Menkhorst, E. McDonald-Madden, and H. Possingham. 2012. "Acting Fast Helps Avoid Extinction." *Conservation Letters*, 5(4): 274-280.

10. Spatz, D., K. Zilliacus, N. Holmes, S. Butchart, P. Genovesi, G. Ceballos, B. Tershy, and D. Croll. 2017. "Globally Threatened Vertebrates on Islands with Invasive Species." *Science Advances*, 3(10): e1603080. DOI: 10.1126/sciadv.1603080.

11. Bellard, C., P. Cassey, and T. Blackburn. 2016. "Alien Species as a Driver of Recent Extinctions." *Biology Letters,* 12(2): 20150623. http://dx.doi.org/10.1098/rsbl.2015.0623.

12. Duhr, M., E. Flint, S. Hunter, R. Taylor, B. Flanders, G. Howald, and D. Norwood. 2019. "Control of House Mice Preying on Adult Albatrosses at Midway Atoll National Wildlife Refuge." In Veitch, C., M. Clout, A. Martin, J. Russel, and C. West (editors), I*sland Invasives: Scaling Up to Meet the Challenge*. IUCN Species Survival Commission Occasional Paper 62, Gland, Switzerland. p. 21–25.

13. Veitch, C.R., M.N. Clout, and D.R. Towns (editors), *Island Invasives: Eradication and Management*. Proceedings of the International Conference on Island Invasives. 2011, Gland, Switzerland and Auckland, New Zealand: IUCN.

14. Wolf, C., H. Young, K. Zilliacus, A. Wegmann, M. McKown, N. Holmes, B. Tershy, R. Dirzo, S. Kropidlowski, and D. Croll. 2018. "Invasive Rat Eradication Strongly Impacts Plant Recruitment on a Tropical Atoll." *PLOS ONE,* 13(7): p. e0200743.

15. Lafferty, K., J. McGlaughlin, D. Gruner, T. Bogar, A. Bui, J. Childress, M. Espinoza, E. Forbes, C. Johnston, M. Klope, K. Miller-Ter, M. Lee, K. Plummer, D. Weber, R. Young, and H. Young. 2018. "Local Extinction of the Asian Tiger Mosquito (*Aedes albopictus*) following Rat Eradication on Palmyra Atoll." *Biology Letters*, 14(2).

16. VanderWerf, E., L. Young, C. Kohley, M. Dalton, R. Fisher, L. Fowlke, S. Donohue, and E. Dittmar. 2019. "Establishing Laysan and Black-Footed Albatross Breeding Colonies Using Translocation and Social Attraction." *Global Ecology and Conservation,* 19: p. e00667.

17. Threatened Island Biodiversity Partners. 2014. *The Threatened Island Biodiversity Database* developed by Island Conservation, University of California Santa Cruz Coastal Conservation Action Lab, BirdLife International and IUCN Invasive Species Specialist Group. Version 2014.1. Available at http://tib.islandconservation.org.

18. Joppa, L., B. O'Connor, P. Visconti, C. Smith, J. Geldmann, M. Hoffman, J. Watson, S. Butchart, M. VirahSwamy, B. Halpern, S. Ahmed, A. Balmford, J. Sutherland, M. Harfoot, C. Hilton-Taylor, W. Foden, E. Di Minin, S. Pagad, P. Genovesi, J. Hutton, and N. Burgess. 2016. "Filling in Biodiversity Threat Gaps." *Science*, 352(6284): p. 416-418.

19. *The Database of Island Invasive Species Eradications* (DIISE). 2014. Developed by Island Conservation, Coastal Conservation Action Laboratory UCSC, IUCN SSC Invasive Species Specialist Group, University of Auckland and Landcare Research New Zealand. http://diise.islandconservation.org.

20. Holmes, N., B. Keitt, D. Spatz, D. Will, S. Hein, J. Russel, P. Genovesi, P. Cowan, and B. Tershy. 2011. "Tracking Invasive Species Eradications on Islands at a Global Scale." In Veitch, C., M. Clout, A. Martin, J. Russel, and C. West (editors), *Island Invasives: Scaling Up to Meet the Challenge*. IUCN Species Survival Commission Occasional Paper 62, Gland, Switzerland. p. 628–632.

21. Spatz, D., K. Newton, R. Heinz, B. Tershy, N. Holmes, S. Butchart, and D. Croll. 2014. "The Biogeography of Globally Threatened Seabirds and Island Conservation Opportunities." *Conservation Biology*, 2014. 28(5): p. 1282-1290.

22. Holmes, N., D. Spatz, S. Oppel, B. Tershy, D. Croll, B. Keitt, P. Genovesi, I. Burfield, D. Will, A. Bond, A. Wegmann, A. Aguirre-Muñoz, A. Raine, C. Knapp, C. Hung, D. Wingate, E. Hagen, F.Méndez-Sánchez, G. Rocamora, H. Yuan, J. Fric, J. Millett, J. Russell, J. Liske-Clark, E. Vidal, H. Jourdan, K. Campbell, K. Springer, K. Swinnerton, L. Gibbons-Decherong, O. Langrand, M. de L. Brooke, M. McMinn, N. Bunbury, N. Oliveira, P. Sposimo, P. Geraldes, P. McClelland, P. Hodum, P. Ryan, R. Borroto-Páez, R. Pierce, R. Griffiths, R. Fisher, R. Wanless, S. Pasachnik, S. Cranwell, T. Micol, and S. Butchart. 2019. "Globally Important Islands Where Eradicating Invasive Mammals Will Benefit Highly Threatened Vertebrates." *PLOS ONE*, 14(3): p. e0212128.

23. Jones, H., N. Holmes, S. Butchart, B. Tershy, P. Kappes, I. Corkery, A. Aguirre-Muñoz, D. Armstrong, E. Bonnaud, A. Burbidge, K. Campbell, F. Courchamp, P. Cowan, R. Cuthbert, S. Ebbert, P. Genovesi, G. Howald, B. Keitt, S. Kress, C. Miskelly, S. Oppel, S. Poncet, M. Rauzon, G. Rocamora, J. Russell, A. Samaniego-Herrera, P. Seddon, D. Spatz, D. Towns, and D. Croll. 2016. "Invasive Mammal Eradication on Islands Results in Substantial Conservation Gains." *Proceedings of the National Academy of Sciences*, 113: p. 4033–4038.

Image credits—pages 4–5: *Tahanea Atoll* courtesy of Island Conservation; page 6: *Matureivavao*, Steve Cranwell, Island Conservation, *Robben Island* and Australia, Wikimedia (Creative Commons); page 7–8: all antiquity maps, Wikimedia (Creative Commons); page 9, *Micronesian Stick Chart*, Walter Meayers Edwards, National Geographic Image Collection; page 17: *Floreana mockingbird*, Tommy Hall, Island Conservation, *Polynesian ground dove*, Marie-Helene Burle/Island Conservation, *Predator damage*, Rory Stansbury/Island Conservation, *Rainbow*, Jessi Behnke/Pacific Rim Conservation; page 20: *Manta rays*, Aurora Alifano/Island Conservation.

The authors appreciate the helpful reviews of David Helweg and Helen Sofaer of the US Geological Survey. Any use of trade, firm, or product names is for descriptive purposes only and does not imply endorsement by the US government.

UNLOCKING OCEAN INTELLIGENCE

With a veritable deluge of new data sources for oceans coming online from satellites, shipboard surveys, and autonomous systems, transforming raw data into meaningful information has emerged as a crucial need for marine industries and management across a broad spectrum of communities. The National Oceanic and Atmospheric Administration and Bureau of Ocean Energy Management have successfully deployed an advanced geographic information systems platform to unleash the power of spatial analytics to unlock ocean intelligence.

By Lisa C. Wickliffe, Seth J. Theuerkauf, Jonathan A. Jossart, Mark A. Finkbeiner, David N. Stein, Christine M. Taylor, Kenneth L. Riley, and James A. Morris, Jr.

The world's largest collection of "ocean intelligence" can now be accessed to help sustain and grow one of the world's largest blue economies.

—Neil Jacobs, PhD, acting administrator,
National Oceanic and Atmospheric Administration (NOAA)

THE OCEAN PLANNING CHALLENGE

With more than 11 million square kilometers of space, oceanic waters of the United States represent one of the largest Exclusive Economic Zones in the world.[1] To manage ecosystem and industry planning decisions in such a vast area, coastal managers increasingly rely on comprehensive geospatial data and information to guide decision-making.[2,3,4,5] The emergence of advanced data acquisition platforms such as satellite and autonomous systems has increased the volume and availability of geospatial data to inform these decisions. However, coastal managers still struggle to turn this data into comprehensive information (i.e., information for decision support) for applications in ocean planning and management.

The US Bureau of Ocean Energy Management (BOEM) and the National Oceanic and Atmospheric Administration (NOAA) partnered to address this challenge and developed OceanReports, an automated geospatial tool for analyzing and visualizing US ocean space. The tool unlocks authoritative ocean planning data to answer essential questions in seconds about planning, regulating, permitting, rulemaking, and efforts toward conserving the diversity of ocean resources and assets in the United States.

Coastal managers tasked with regulating ocean space and the industries occupying it require the best available information to make confident decisions regarding current and future ocean uses. Ocean planning requires coastal managers to consider the environmental diversity and array of uses throughout our oceans—ranging from the distribution of sensitive habitats to the prevalence of vessel traffic—and minimize conflicts among them. Science-based geospatial tools allow them to address specific ocean management challenges and advance economic development and conservation goals.

Block Island Wind Farm offshore from Rhode Island in the Atlantic Ocean.

The Port of Oakland in Northern California is typical among major US ports in terms of the demands placed on it from various stakeholders, including shipping, fishing, the cruise industry, and the general public, to name a few.

The multiple dimensions of the ocean space

Some of the major goals of ocean planning include restoring infrastructure, protecting critical habitats such as migratory corridors for endangered species, and managing ecosystems. For ocean industry, marine planning includes matching the most appropriate ocean use to an ocean space. Oceans within US jurisdiction play host to numerous military and industrial objectives (e.g., energy production, major communication hubs, movement of goods and services, training for military readiness), as well as fisheries, recreation, and tourism activities. These activities occur in the shared ocean space with diverse marine habitats, sensitive species, and many other natural resources. Within the ocean planning schema for new activities, coastal planners must pay close attention to the characterization of the ocean neighborhood in which an ocean activity will occur over space and time.

Planners require spatial data at multiple scales and dimensions to better understand how different activities share a common ocean footprint. Most maps are 2D and may leave the false impression that the ocean contains virtually no unused space. In reality, information needed for ocean planning spans multiple dimensions, including the seafloor, water column, sea surface, and interactions with the atmosphere above the ocean. Consideration of all dimensions is essential for conflict avoidance (e.g., shipping traffic, sensitive habitats). Further, some data must be considered over time, as some datasets only apply to certain seasons or time of day (e.g., migration of whale species, fish spawning aggregations).

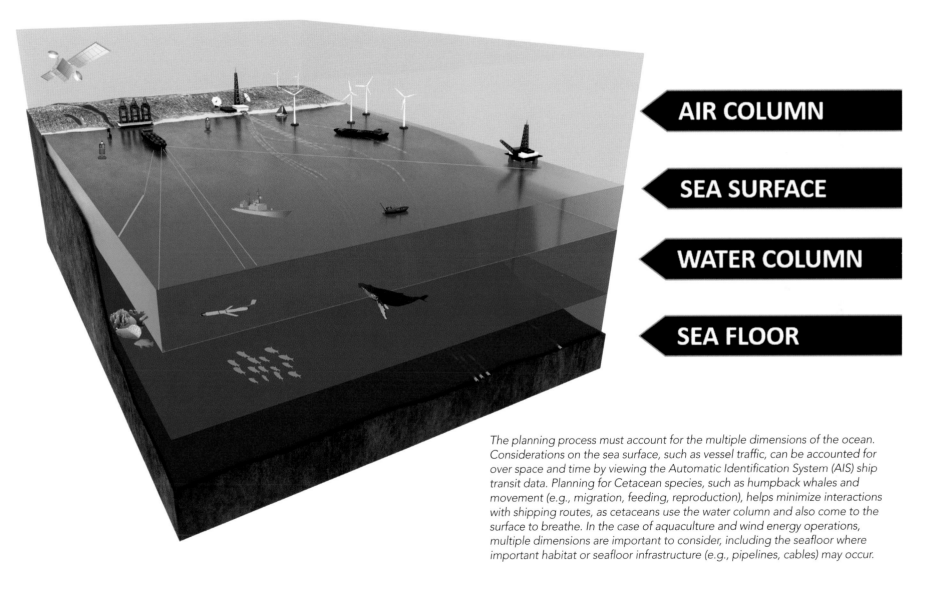

AIR COLUMN

SEA SURFACE

WATER COLUMN

SEA FLOOR

The planning process must account for the multiple dimensions of the ocean. Considerations on the sea surface, such as vessel traffic, can be accounted for over space and time by viewing the Automatic Identification System (AIS) ship transit data. Planning for Cetacean species, such as humpback whales and movement (e.g., migration, feeding, reproduction), helps minimize interactions with shipping routes, as cetaceans use the water column and also come to the surface to breathe. In the case of aquaculture and wind energy operations, multiple dimensions are important to consider, including the seafloor where important habitat or seafloor infrastructure (e.g., pipelines, cables) may occur.

OCEAN REPORTS: AUTOMATING OCEAN PLANNING ANALYSIS

MarineCadastre.gov, a cooperative effort by BOEM and NOAA, provides authoritative ocean data, tools, and support to marine planning communities. The website organizes marine planning data and disseminates it into the public domain. At its core, MarineCadastre.gov contains data on jurisdictional boundaries, marine infrastructure, transportation, alternative energy, traditional energy, physical factors, and biological data to support planning, management, and conservation of marine spaces.

OceanReports began in 2014 as a prototype to help ocean planners, industry representatives, and regulators more easily query a specific area of ocean space and receive essential summarized information. Available in seconds in the form of graphics and statistics, the information is used to inform planning decisions. Through unlocking spatial data and analysis, OceanReports increases the power and utility of data for technical and nontechnical users such as coastal managers, environment-focused nongovernmental organizations (eNGOs), environmental policy analysts, geographic information systems (GIS) managers, K–12 educators, international partners, industry consultants, and congressional and policy staff.

Designed as a freely available web application, OceanReports allows users with no technical experience in GIS to select an area of US ocean space and instantaneously obtain more than 80 unique infographics derived from an automated spatial analysis of data associated with that location. These infographics include information on energy and minerals, natural resources, transportation and infrastructure, the oceanographic and biophysical conditions, and the local ocean economy. For anywhere in US ocean waters—from the coastal shelf of Florida to the Bering Sea of

Alaska to the far ocean reaches of the Pacific Islands—users can start with an area of ocean space in mind and in return receive a comprehensive automated report detailing key environmental and space use considerations essential for planning, as shown in the map depicting harmful algae bloom data off Florida's Gulf Coast.

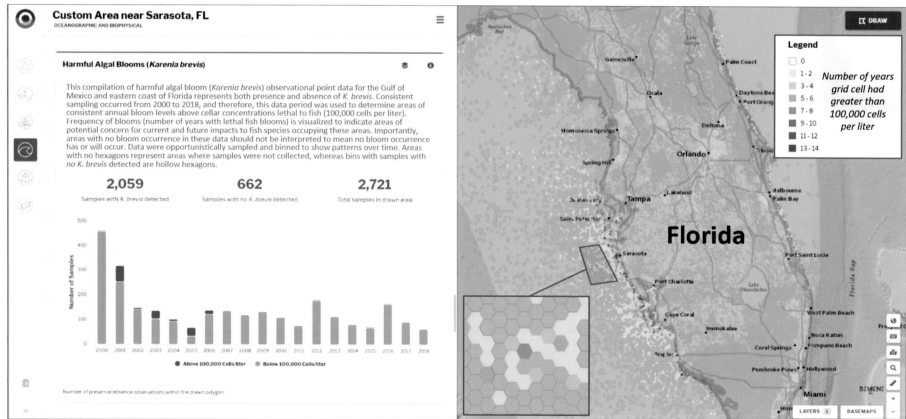

Coastal managers can infer biological risks in a given area by visualizing harmful algae bloom (HAB)—in this case , Karenia brevis near Sarasota, Florida—in terms of its occurrence and frequency over time at cellular levels harmful to finfish. This information applies to fisheries management and inland and offshore aquaculture screening and siting.

OceanReports themes and data

OceanReports delivers a customized report for a user-derived ocean "area of interest" that analyzes and distills key spatial data and provides location-specific insights in these six topic areas:

1. **General information**: Describes the size of the selected area, jurisdictional and political boundaries, land elevation and water depth, relevant laws, and populated places.

2. **Energy and minerals**: Highlights the energy potential available off the coast of the United States, current areas of extraction/collection, and the availability of mineral resources used to restore hundreds of miles of coastline and protect billions of dollars in infrastructure and ecological habitat from coastal erosion and destructive storms.

3. **Natural resources and conservation**: Provides information on the distribution and abundance of natural resources (e.g., habitat locations that support migratory and endangered species). These considerations are essential for balancing a healthy economy and coastal ecosystems through mitigating impacts to natural resources.

4. **Oceanographic and biophysical**: Offers critical information on a variety of oceanographic parameters, including data resources derived from the Ecological Marine Unit.

5. **Transportation and infrastructure**: Shows the infrastructure and activities of the marine transportation sector along the US coastline, including information on vessel traffic, routing, and restricted areas. This topic area also provides information about permanent and semi-permanent structures such as cables, pipelines, ports, oil platforms, and wells.

6. **Economics and commerce**: Provides information on relevant marine-dependent jobs, commercial fish landings, census-derived indicators, and other key economic considerations.

Chapter Theme	Symbol	Infographic Reports Available		
General Information		Report Area Depth/Elevation Populated Places Federal/State/County Jurisdictions	Congressional and Legislative Districts Federal Statutes Indian Land Areas	
Energy & Minerals		Offshore Wind Potential Offshore Wind Planning Areas Offshore Wind Energy Leases Oil and Gas Potential	Oil and Gas Planning Areas Oil and Gas Leases Coastal Energy Facilities OCS Blocks with Sand Resources Federal Sand and Gravel Leases	Beach Nourishment Projects Surficial Sediment Texture Ocean Disposal Sites
Transportation & Infrastructure		AIS Vessel Count Vessel Routing N. Atlantic Right Whale Management Areas Anchorage Areas	Ports Coastal Maintained Channels Danger Zones/Restricted Areas Unexploded Ordnances Formerly Used Defense Sites	Wrecks/Obstructions Cables and Pipelines Wastewater Outfalls Aquaculture Oil Lightering Zones Deepwater Ports Oil/Gas Platforms Oil/Gas Wells Pilot Boarding Areas
Natural Resources		Endangered Species ESA-Critical Habitat Designations Managed Highly Migratory Species Audubon Important Bird Areas Coastal Barrier Resource Areas	Protected Areas Artificial Reefs Shallow Corals Deep-sea Sponge/Coral Obs. Deep-sea Coral Habitat Suitability	Historical Lighthouses Cetacean Biologically Important Areas Essential Fish Habitat
Oceanographic & Biophysical		Wave Height, Period and Direction Wind Speed and Direction Current Speed and Direction Sea Surface Height Water Temp/Salinity at depth	Nitrate concentration Phosphate concentration Silicates concentration Aragonite Saturation State Light Attenuation (Kd PAR)	Light Attenuation (Kd 490) Chlorophyll a Concentration Harmful Algae Blooms (K. brevis) Historical Tropical Cyclone Exposure
Economics & Commerce		Ocean Job Contributions GDP of Ocean Economy Contributions by Sector	Census Statistics Fishing Economic Value (North and Mid Atlantic) Commercial Fish Landings	

OceanReports chapter themes, the symbol representing that theme, and the infographics present in each theme where statistics are provided for the user.

One of the most important aspects of OceanReports is the built-in ability for users to immediately access the data underpinning the tool through map layer viewing, available downloads, and the original data source information provided in the metadata. This example from OceanReports shows a report for the lower Chesapeake Bay in Virginia state waters and a snapshot of the economics and commerce for that area.

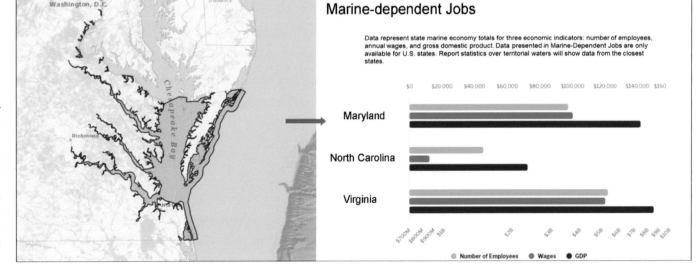

Infographics presentation

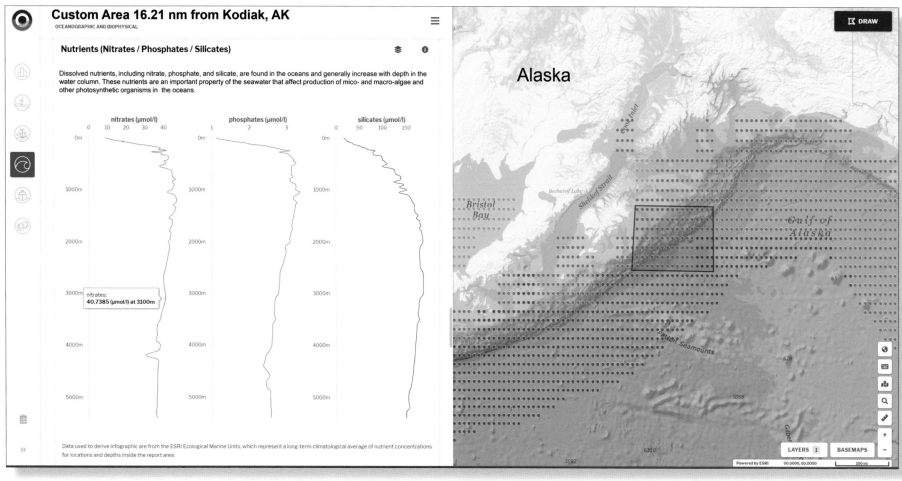

Custom Area 16.21 nm from Kodiak, AK
OCEANOGRAPHIC AND BIOPHYSICAL

Nutrients (Nitrates / Phosphates / Silicates)

Dissolved nutrients, including nitrate, phosphate, and silicate, are found in the oceans and generally increase with depth in the water column. These nutrients are an important property of the seawater that affect production of mico- and macro-algae and other photosynthetic organisms in the oceans.

nitrates:
40.7385 (µmol/l) at 3100m

Data used to derive infographic are from the ESRI Ecological Marine Units, which represent a long-term climatological average of nutrient concentrations for locations and depths inside the report area.

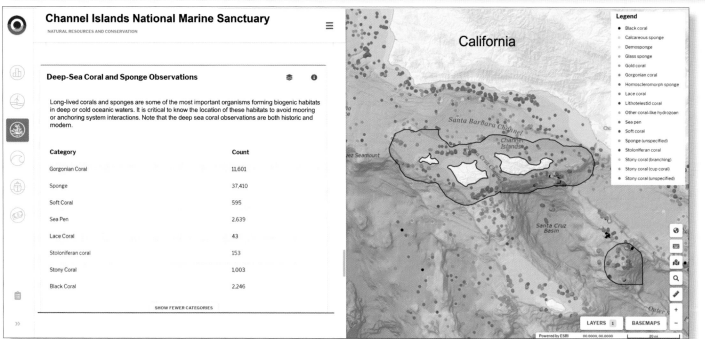

Channel Islands National Marine Sanctuary
NATURAL RESOURCES AND CONSERVATION

Deep-Sea Coral and Sponge Observations

Long-lived corals and sponges are some of the most important organisms forming biogenic habitats in deep or cold oceanic waters. It is critical to know the location of these habitats to avoid mooring or anchoring system interactions. Note that the deep sea coral observations are both historic and modern.

Category	Count
Gorgonian Coral	11,601
Sponge	37,410
Soft Coral	595
Sea Pen	2,639
Lace Coral	43
Stoloniferan coral	153
Stony Coral	1,003
Black Coral	2,246

SHOW FEWER CATEGORIES

Legend
- Black coral
- Calcareous sponge
- Demosponge
- Glass sponge
- Gold coral
- Gorgonian coral
- Homoscleromorph sponge
- Lace coral
- Lithotelestid coral
- Other coral-like hydrozoan
- Sea pen
- Soft coral
- Sponge (unspecified)
- Stoloniferan coral
- Stony coral (branching)
- Stony coral (cup coral)
- Stony coral (unspecified)

Sampling of infographics generated by OceanReports for the Ecological Marine Units (EMU) nutrient concentrations at depth in the Gulf of Alaska (top) and deep sea coral species in Southern California. Each infographic is built from spatial analysis of underlying data to provide customized key statistics and information for the user-selected area of interest.

Data exploring via the map viewer

Beyond the distilled information that OceanReports provides users for a given area of ocean space, the tool allows users to explore spatial data within the integrated map viewer. This capability allows deeper engagement and interaction, depending on user interest. Users can input specific coordinates to search for a location of interest, adjust underlying basemaps, and print and share reports, as shown. Each infographic in the six themes contains these functional components:

A   Home button to return user to start page
B   Themed chapters
C   Access to Quick Reports
D   Begin drawing a custom area or view Quick Reports options
E   View regions, input coordinates, search, measure, change basemap
F   Switch layers on a map or look at industry themes

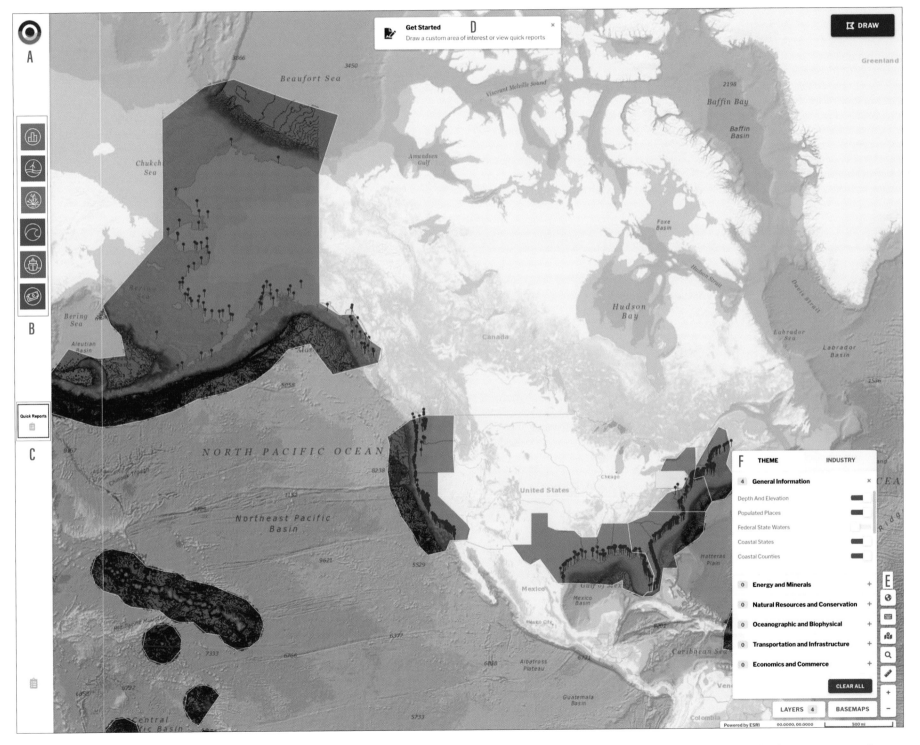

Under the hood

OceanReports provides synthesized and authoritative information early in the permitting or planning process in an effort to provide a "first look" at ocean interactions, with the goal of long-term ecosystem sustainability in mind. Providing this information early in these processes can greatly impact the project trajectory and outcome. A comprehensive ocean neighborhood (i.e., a geographically localized community that interacts or overlaps with a proposed site) analysis allows decision-makers to see the big picture and make screening-level decisions quickly and more efficiently. It also increases transparency in decision-making by allowing all stakeholders, from developers to regulators, to use the same data and information to make decisions. OceanReports ultimately streamlines communication between the public, agency, and industry; agency to agency; and agency to the public.

Users can come to OceanReports with a predefined planning goal prior to exploring an ocean area of interest (i.e., custom area). Custom areas can be drawn in the tool, representing ocean areas of interest given a user's specific geographic focus. The back-end software then accesses the necessary data associated with that custom area and applies specified reporting rules that define the automated spatial analysis (e.g., report what is inside or intersects with the custom area); then parallel automatic processing generates the themed infographic report for the custom ocean area. In some cases, maximum area thresholds limit the descriptive statistics returned for an area that is too large to provide valid statistics. These were determined for each continuous data layer using the Moran's I spatial dependence test. [6]

Although visualizing mapped data can provide useful information to guide decisions, summary statistics derived from the underlying spatial data can provide deeper insights. Subject matter experts in specific coastal planning topics such as wind, oil and gas, marine minerals and aquaculture, helped develop rules to guide the return of statistical and attribute information within each infographic. Each infographic follows a rule specific to the data being analyzed and returns a result for the user-defined custom area. To provide useful and intuitive statistics and graphical summaries of data for each infographic in OceanReports, numerous graphical displays were developed to convey easily interpreted important information about a user's specific ocean area of interest (i.e., custom area). Infographics complement each dataset displayed on the map viewer portion of a custom report and offer critical insights about the user-defined area. Examples of graphical depictions of data for infographics displayed for a custom area include interactive tables and charts, rose plots for oceanographic variables, monthly displays of data with temporal components, and profiles to illustrate data at various depth levels.

OceanReports Process & Benefits

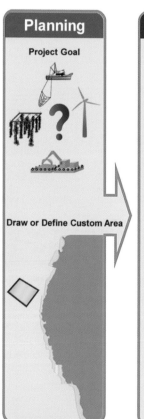

Planning
Project Goal

Draw or Define Custom Area

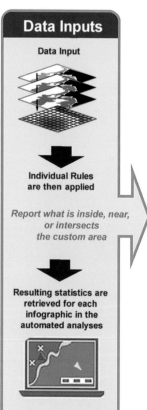

Data Inputs
Data Input

Individual Rules are then applied

Report what is inside, near, or intersects the custom area

Resulting statistics are retrieved for each infographic in the automated analyses

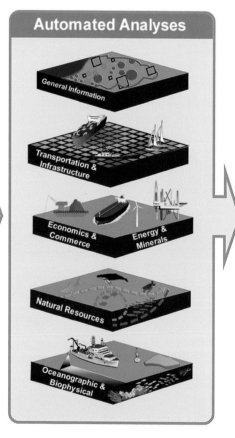

Automated Analyses
General Information

Transportation & Infrastructure

Economics & Commerce

Energy & Minerals

Natural Resources

Oceanographic & Biophysical

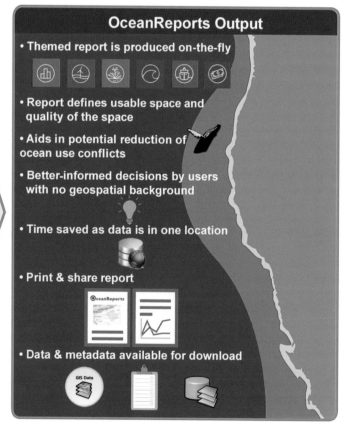

OceanReports Output
- Themed report is produced on-the-fly
- Report defines usable space and quality of the space
- Aids in potential reduction of ocean use conflicts
- Better-informed decisions by users with no geospatial background
- Time saved as data is in one location
- Print & share report
- Data & metadata available for download

BOEM

The nuts and bolts

The basic structure of a web mapping application includes the front end (client) and the back end (server). The user interacts with the front end, most commonly in a web browser window. Behind the scenes, the application pulls, processes, returns, and displays data to the user.

A cloud server hosts OceanReports using Microsoft Azure™ Platform as a Service (PaaS). The application uses the Google-developed Angular web application framework. The application also needs additional open source libraries and dependencies that are included using a JavaScript module bundler called Webpack. For the presentation of the application in a web browser, styles are added using a powerful cascading style sheets (CSS) preprocessor, known as SASS. The Highcharts JavaScript charting library powers the charts featured in many infographics in the application. A Socket.io connection establishes real-time communication between the server and client, allowing data to be updated quickly in the custom report infographics. Atlassian's Bitbucket and Bamboo provide continuous integration and deployment, ensuring each line of code is source controlled, QA tested, and deployed automatically to the cloud servers.

The two main components that comprise the back end of the application are a runtime environment and a database. The runtime environment uses Express, a NodeJS server framework, which handles the dynamic calculations, conditional statements, and functions for the application. An Azure database for PostgreSQL allows for fast indexing and querying of the 86 vector and 72 raster data layers (i.e., relational database), and the application uses PostGIS for storing spatial datasets. Esri's ArcGIS Map Services display and render map layers stored within the Azure database. Because of the complexity of the data and large load requests, the application uses Nginx for load balancing and a fast reverse proxy.

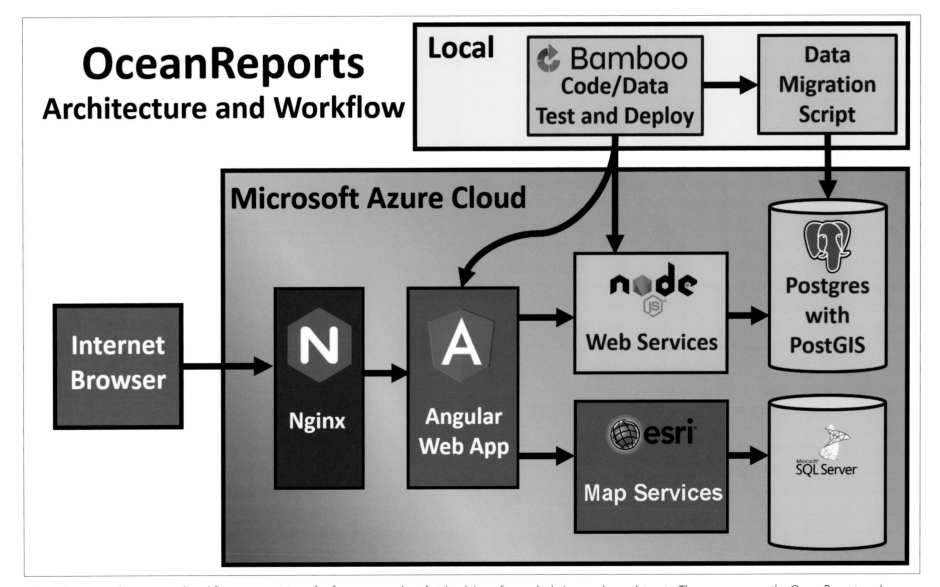

OceanReports architecture and workflow uses a variety of software to produce fast load times from calculations on large datasets. The user accesses the OceanReports web application from an internet browser, while in the background the application accesses several software and service platforms to produce and retrieve the user's request. Local machines are used to update and improve the web application.

SITING A WIND FARM UNDER BOEM'S RENEWABLE ENERGY PROGRAM

BOEM is the federal agency in charge of planning and leasing renewable energy in US ocean waters. The BOEM Office of Renewable Energy Programs facilitates responsible development of renewable energy on the outer continental shelf through conscientious planning, stakeholder engagement, comprehensive environmental analysis, and sound technical review.[8] The first step identifies potential wind-planning areas by the state or region's renewable energy task force. A renewable energy task force is created when a state or a region expresses interest in the development of renewable energy in federally managed US waters and is made up of representatives from the states, federally recognized tribes, and federal agencies that have an interest in the area. A task force considers numerous factors in choosing a site for an offshore wind farm. These considerations include costs, potential profits, available resources, potential conflicts, and political will. Considerations for design and engineering of the structure depend on site-specific conditions, particularly water depth, seabed geology, and wave loading.[8] In choosing a location, the task force recommends potential lessees to consider where the energy is needed and whether the wind farm is close enough to supply those needs.

All new wind energy farms must comply with BOEM's renewable energy program, which occurs in four phases: 1) planning and analysis, 2) lease issuance, 3) site assessment, and 4) construction and operations. The planning and analysis phase seeks to identify suitable areas for wind energy leasing consideration through collaborative, consultative, and analytical processes that engage stakeholders, tribes, and state and federal government agencies. During this phase, BOEM reviews environmental compliance and consults with tribes, states, and natural resource agencies (https://www.boem.gov/Commercial-Leasing-Process-Fact-Sheet).

Before or during this phase, potential stakeholders in the region often start with a general area of interest. Using OceanReports, we can begin to guide some decisions that allow the task force to view the same general area or areas of interest. Stakeholders should consider answering these kinds of questions before beginning the first phase:

- Where is the energy needed? Is there a potential location with good wind resources close to energy needs?
- Is the wind strong enough year-round to keep the turbines going at a profitable rate?
- Is the bottom type appropriate for a wind turbine installation?
- Is it shallow enough (under 100 feet) to support a seafloor-supported structure or is it greater than 100 feet, suggesting a floating wind farm?
- Are there ports and transmission connections that can handle the construction and operational needs?
- Which authorities can approve, permit, lease, monitor, and evaluate the project?
- What are the regulations that must be considered?
- What other human activities regularly occur in the area?
- Are there environmental issues (wildlife harassment, air pollution, water pollution, potential effects to Essential Fish Habitat areas, etc.)?

In this hypothetical example, stakeholders wanted to locate a wind farm off Montauk, New York. The investor group and the state are interested in the area shown on the map off Long Island, as these stakeholders operated a successful wind farm within Rhode Island state waters (i.e., Block Island Wind Farm). The investor group wanted to move to the open ocean that lies in federally managed waters, as offshore winds tend to blow harder and more uniformly than on land, and thus can produce significantly more energy/electricity.[8] Additionally, siting turbines farther offshore makes them less visible from shore, minimizing potential visual conflicts. Nearby, ports and transmission connections are already in place that can handle construction

Artist rendering of proposed offshore wind turbine designs gives stakeholders and the public a sense of the visual impact of massive renewable energy projects.

and operational needs. To further investigate the proposed area, the investor group used OceanReports to inspect several other questions that need consideration, including depth, other human ocean activities in the area, characterization of the surrounding ecosystem, investigation of sensitive habitat and species, bottom type, and the potential for wind energy in the area of interest.

First, the investors drew a custom area within the OceanReports tool for the area of interest for siting the wind farm (next page). Next, the investor group reviewed each theme chapter within OceanReports for the custom area, which provides synthesized information about the drawn area. The general information theme provided the general characteristics, including the size of the area (281 km²), minimum (−37.2 m) and maximum depth (−60.6 m), and whether the area is in federal waters or state waters. In this case, the investor group aimed to be in federal waters, and 100 percent of the custom area was in federal waters.

Next, the group checked wind energy potential and found it was "outstanding" for the custom area. Looking further into the custom area, the group checked for substrate type because it is one of the main factors in determining whether and what type of turbines should be used. By looking at the Energy and Minerals theme information, the group saw the predominant substrate in the area is sand, which is favorable for the project. Next, clicking the Natural Resources theme, the group saw that 100 percent of the area is in North Atlantic Right Whale Seasonal Management Area, which dictates caution for ship traffic in migration season. The Cetacean Biologically Important Area Fin Whale feeding zone covers part of the area, and the Northern Right Whale Migration Area covers all of the area.

At this point in the automated analysis, the investor group considered reviewing the biological opinion (i.e., formal consultation, stating the opinion of the agency on whether a federal action is likely to jeopardize the continued existence of listed species or result in the destruction or adverse modification of critical habitat) if they proceed from the nearest wind energy lease blocks.[9] Clicking the Transportation and Infrastructure theme showed that vessel traffic is relatively low, but seafloor infrastructure may impede progress, despite the three electric power facilities nearby. Submarine cables run through the proposed wind farm area, but other proposed sites have submarine cables running in subsurface sediment. Because of the uncertainty associated with potential interaction with the cables, the group then moved the custom area by dragging the original drawn area to a nearby adjacent area to avoid the cables. After turning on unexploded ordnance and shipwrecks, the group decided to go farther offshore to avoid the aforementioned constraints. The project was still economically viable but would require a different and more experimental type of turbine platform such as tripoled, jacketed, or tripod wind turbine structures in the deeper waters. After checking several oceanographic factors (e.g., tropical cyclone wind exposure, significant wave height, prevailing wind direction), the investor group decided to pursue further conversations with relevant agencies to determine whether the location is truly viable and permissible.

In real-world applications of OceanReports, the custom area would be drawn multiple times within multiple areas, allowing the user to run a report for each area to share with collaborators and state and federal government agencies and compare the options for each area against one another. Not all intersecting data layers become roadblocks to development. Wind farms can build around cables to avoid them; protected areas for whales can be utilized as long as construction doesn't occur while they are in the area. Finding just the right area that eliminates all conflict is nearly impossible, but avoiding already known impediments before progressing further in the planning process can save significant time. OceanReports helps users quickly identify the known challenges for each area, understand needed further investigation, and can use finer-scale data and information to provide needed detail once a potential area is identified.

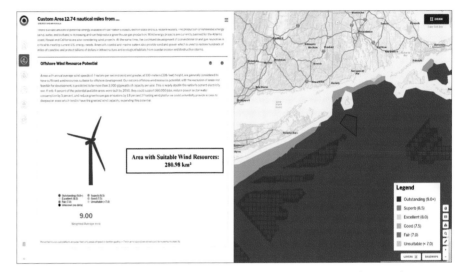

Hypothetical custom area drawn off the coast of Montauk, New York, to determine whether the Custom Area characteristics meet the needs of the proposed wind farm.

Wind energy potential for the drawn custom area. Here, we can see the area has "outstanding" potential and does not directly overlap with known lease areas.

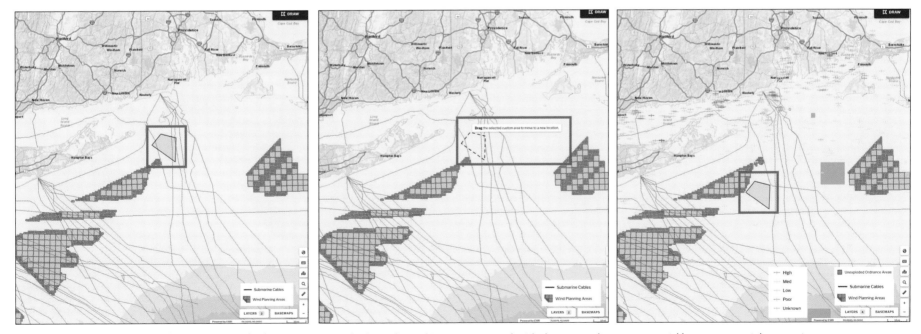

After reviewing the seafloor infrastructure inside the custom area, the hypothetical investor group decided to move the area to avoid known potential constraints.

PIONEERING OFFSHORE AQUACULTURE IN THE GULF OF MEXICO

Aquaculture is one of the fastest-growing food production sectors in the world, and it plays an increasing role in sustainable seafood production across the United States. The need to increase food security, reduce a multibillion-dollar seafood trade deficit, create jobs, and revitalize coastal communities drives aquaculture development in the United States. Many regions around the country are preparing for this increase as aquaculture aims to move offshore. The Gulf of Mexico offers substantial potential for development of marine aquaculture. During the past decade, industry and proponents for aquaculture have tried to develop frames of reference and rationales for creating an offshore aquaculture industry in the Gulf. Stakeholders and residents in coastal communities have intensely debated aquaculture development. Coastal managers and stakeholders need awareness and confidence to use science-based decision tools to inform coastal ocean use plans and equitably resolve points of resistance to industry development.

Although the United States owns immense ocean space, identifying suitable locations for commercial aquaculture development requires specific environmental conditions and must minimize conflict with natural resources such as sensitive habitats, as well as established ocean industries such as energy, mining and mineral extraction, recreational and commercial fishing, navigation, military, shipping, and other public interests. The task of identifying these locations is particularly important within the Gulf of Mexico, where abundant resources have contributed to a growing ocean economy. For instance, more than 90 percent of US oil and gas production occurs in the Gulf of Mexico, providing billions of dollars to the national economy. The multibillion-dollar shipping and shipbuilding industries include two of the largest ports in the world, Houston and New Orleans. Gulf fisheries are some of the most productive in the world and yield more finfish, shrimp, and shellfish annually than the South and Mid-Atlantic, Chesapeake Bay, and New England areas combined. The region is home to three of the top eight fishing ports in the nation by weight, and five of the top 20 fishing ports in the nation by dollar value.[10]

In what has been perceived as a sea of conflict, OceanReports lends support to regulators and industry in prospecting for suitable locations for aquaculture development. Locating an aquaculture operation offshore is an expensive endeavor. Siting and reconnaissance in the unprotected open ocean environment depend on using the best available science to account for increased exposure to extreme weather and ocean conditions, competition for space, and protection of natural resources. In this hypothetical example, industry and academic partners in the Gulf region want to deploy a new, technologically advanced finfish farm in tandem with a decommissioned oil platform. The Energy Policy Act of 2005 granted BOEM jurisdiction over projects that use existing (decommissioned) oil and natural gas platforms for other purposes in federal waters, in addition to jurisdiction over renewable energy projects. Alternate uses of existing facilities may include, but are not limited to, research, education, offshore aquaculture, support for offshore operations and facilities, and telecommunication facilities.[8] These infrastructures can withstand high-energy systems and might eventually reach economies of scale that offset some of the additional costs of offshore locations.[11]

After determining viable decommissioned platforms within the region, the team needed to explore Gulf waters to determine a platform area where oceanic conditions are conducive for finfish aquaculture. In this hypothetical example, the team aims to moor the finfish cage near the rigid platform structure. The team plans to use the structure for worker housing, storage, feed automation, extra parts for cage repair, and as a communication hub to transmit real-time data gathered by sensor systems on the cage to land-based facilities. Given the species of fish for culture, the design

In Hawaiian waters, aquaculture uses the Aquapod finfish cage to grow fish in the open ocean environment.

characteristics of the cage, and the frequency of natural disasters in the region, the group used the automated spatial analytics in OceanReports to quickly screen large areas of ocean space. The analytics inform the basic engineering criteria for culture systems (depth, current speed, sediment type) and environmental conditions required for aquatic species (temperature, water quality). Because of significant wave heights during storms in the Gulf of Mexico, the aquaculture cage must be located in at least 50 meters of water. The group switched on the depth contours to determine this depth before drawing a custom area. Once the depth constraint was met, the team investigated the Oceanographic and Biophysical theme for significant wave height and sediment type because these variables also limit gear type and placement. The team assessed temperature and salinity at the depth for culture species within the drawn custom area to determine whether the candidate species of finfish could thrive. Further, the team assessed the speed of the current on and below the surface to ensure it never rose above 1 meter per second on average (i.e., faster than the gear or species could sustain over time). During episodic storm events, sensors will alert the team to conditions so they can sink the cage to potentially mitigate damage and losses.

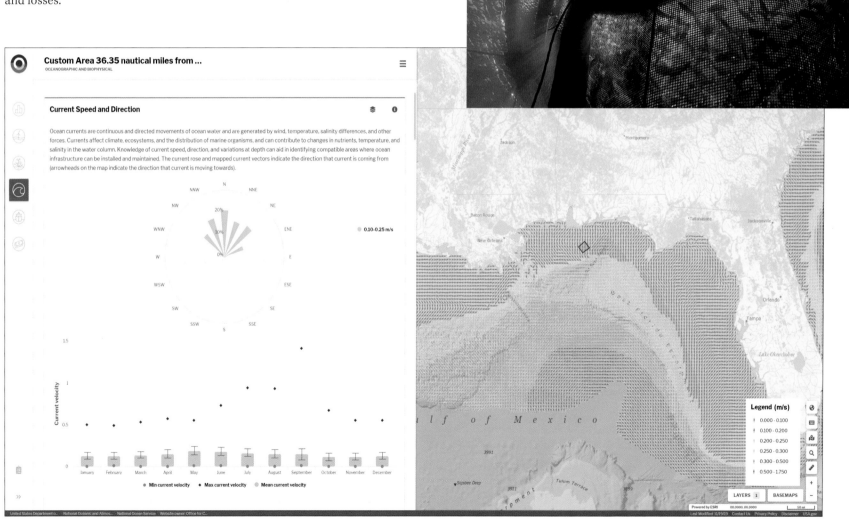

The map portion of OceanReports shows the average current speed at the surface of the ocean during a 20-year climatology, distilled into monthly intervals. The current rose diagram at the top of the infographic gives the prevailing current direction and speeds over the climatology assessed. The bottom infographic depicts monthly maximum, mean, minimum, and standard deviation to determine, for instance, which month the maximum current speeds occur.

As a screening tool, OceanReports gauges opportunities for aquaculture at a regional scale. Generally, a team will find multiple alternative sites before conversations begin with state and federal government permitting agencies, because each site has different sets of constraints that require negotiation. The team can streamline the permitting process by holding prepermitting meetings with agencies and finding several locations that meet required parameters and avoid major ocean use conflicts. OceanReports can screen large ocean spaces to identify major conflicts in an area. The use of high-resolution oceanographic and biophysical data can further define an area and is recommended before farm planning proceeds to the next step.

A conceptual diagram of the finfish cages in the ocean using the rigid structure of the decommissioned oil rig as an automated feeding platform, housing unit for workers, and real-time communication hub to relay any issues to land-based operations.

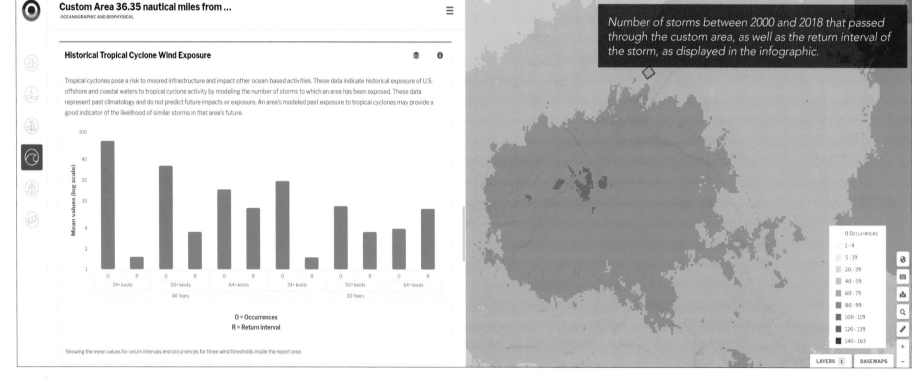

Number of storms between 2000 and 2018 that passed through the custom area, as well as the return interval of the storm, as displayed in the infographic.

Conclusions and insights

OceanReports represents the next chapter in marine spatial planning and a fundamental advance in our ability to access and transform big ocean data. In unlocking access to an unprecedented amount of essential information, OceanReports shares information transparently to empower ocean planning decisions. Public benefits derived from OceanReports include increased regulatory confidence, decreased time and increased efficiency in the permitting process, transparency, and better-informed stakeholders (e.g., public, industry, government, NGOs). Standard web mapping applications or data portals allow users to view and download numerous spatial datasets. OceanReports takes the standard web map application to the next level by allowing the user to draw and define a custom area and receive a customized report detailing essential information for an area of ocean space derived from an automated spatial analysis, usually in less than 2 seconds.

OceanReports helps users engage openly with coastal communities, develop pilot-scale demonstrations, and plan large commercial projects. OceanReports stimulates an objective and rigorous analysis of ocean space to inform planning and preliminary permitting discussions with regulators and also provides transparency to the planning process. Early and informed engagement with the regulatory community—facilitated partly through the information provided via OceanReports—can illuminate crucial trade-offs, incorporate public values and concerns, and explore approaches to minimize environmental impacts.

Author affiliations

Lisa C. Wickliffe, CSS under contract to NOAA/NOS/NCCOS; Seth J. Theuerkauf, NOAA National Marine Fisheries Service, Office of Aquaculture; Jonathan A. Jossart, CSS under contract to NOAA/NOS/NCCOS; Christine M. Taylor, Office of Strategic Resources, Bureau of Ocean & Energy Management; Mark A. Finkbeiner, NOAA National Ocean Service, Office for Coastal Management; Dave N. Stein, NOAA National Ocean Service, Office for Coastal Management; Kenneth L. Riley, NOAA National Ocean Service National Centers for Coastal Ocean Science; and James A. Morris, Jr., NOAA National Ocean Service National Centers for Coastal Ocean Science.

Image credits

Port of Oakland photo by Basil D. Soufi (Wikimedia Creative Commons).

Block Island Wind Farm photo by Ionna22 (Wikimedia Creative Commons).

Offshore wind turbine designs illustration (credit: Josh Bauer, NREL).

Aquapod finfish cage photo by Bryce Groark.

HDPE finfish net pen (courtesy of NOAA Fisheries).

Conceptual diagram of the finfish cages in the ocean using a rigid structure (courtesy of Nippon Steel Engineering).

Some graphic elements are courtesy of the Integration and Application Network, University of Maryland Center for Environmental Science (ian.umces.edu/symbols).

NOTES

1. National Oceanic and Atmospheric Administration (NOAA) General Counsel. 2011. "The United States Is an Ocean Nation." Accessed March 9, 2020. https://www.gc.noaa.gov/documents/2011/012711_gcil_maritime_eez_map.pdf.

2. Good, J.W., and D. Sowers. 1999. "Benefits of Geographic Information Systems for State and Regional Ocean Management." Final Report to the Coastal Services Center, National Oceanic and Atmospheric Administration. *Sea Grant Special Report* 99-01. Corvallis: Oregon State University.

3. Foley, M.M., M.H. Armsby, E.E. Prahler, M.R. Caldwell, A.L. Erickson, J.N. Kittinger, L.B. Crowder, and P.S. Levin. 2013. "Improving Ocean Management through the Use of Ecological Principles and Integrated Ecosystem Assessments." *BioScience*, 63(8):619-631.4.

4. Wright, D.J. (Ed.). 2015. *Ocean Solutions, Earth Solutions*, Redlands, CA: Esri Press, 366 pp., doi: 10.17128/9781589483651.

5. Maxwell, S.M., E.L. Hazen, R.L. Lewison, D.C. Dunn, H. Bailey, S.J. Bograd, D.K. Briscoe, S. Fossette, A.J. Hobday, M. Bennettt, S. Benson, M.R. Caldwell, D.P. Costa, H. Dewar, T. Eguchi, L. Hazen, S. Kohin, T. Sippel, and L.B. Crowder. 2015. "Dynamic Ocean Management: Defining and Conceptualizing Real-Time Management of the Ocean." *Marine Policy*, 58:42-50.

6. Sayre, R., J. Dangermond, D. Wright, S. Breyer, K. Butler, K. Van Graafeiland, M.J. Costello, P. Harris, K. Goodin, M. Kavanaugh, N. Cressie, J. Guinotte, Z. Basher, P. Halpin, M. Monaco, P. Aniello, C. Frye, D. Stephens, P. Valentine, J. Smith, R. Smith, D.P. VanSistine, J. Cress, H. Warner, C. Brown, J. Steffenson, D. Cribbs, B. Van Esch, D. Hopkins, G. Noll, S. Kopp, and C. Convis. 2017. "A New Map of Global Ecological Marine Units–an Environmental Stratification Approach." Washington, DC: American Association of Geographers. 36 pages.

7. Jossart, J., S.J. Theuerkauf, L.C. Wickliffe, and J.A. Morris, Jr. 2020. "Practical Applications of Spatial Autocorrelation Analyses to Improve Marine Aquaculture Siting." Frontiers in Marine Science Special Edition: *Aquaculture and Remote Sensing*, (6)806:1-15. https://doi.org/10.3389/fmars.2019.00806 .

8. Bureau of Ocean and Energy Management (BOEM). 2019. "Renewable Energy on the Outer Continental Shelf." Accessed November 29, 2019. https://www.boem.gov/renewable-energy/renewable-energy-program-overview.

9. US Fish and Wildlife Service (USFWS). 2019. Accessed December 4, 2019. "Endangered Species Glossary." https://www.fws.gov/midwest/endangered/glossary/index.html.

10. National Marine Fisheries Service (NMFS). 2018. Fisheries of the United States, 2017. US Department of Commerce, NOAA Current Fishery Statistics No. 2017 Available at: https://www.fisheries.noaa.gov/feature-story/fisheries-united-states-2017.

11. California Environmental Associates (CEA). 2018. "Offshore Finfish Aquaculture." Global Review and US Prospects. Accessed on: December 3, 2019. https://www.ceaconsulting.com/wp-content/uploads/CEA-Offshore-Aquaculture-Report-2018.pdf

THE GEOGRAPHY OF OCEAN PLASTICS

Since their invention in the 1950s, plastics have had an alarming and highly visible impact on the world's oceans that humanity certainly never anticipated. Modern scientific detectives are turning to big data and advanced GIS software to understand the major sources of plastic pollution in the world's oceans as a first step to reducing their presence.

By Orhun Aydin and Shaun Walbridge, **Esri**

Plastics washed ashore in Accra Beach, Ghana. Of all plastics ever produced, 99.9 percent still exist in their original shape (including discarded plastics).

THE PROBLEM WITH PLASTIC

Understanding the exposure that marine animals face when they encounter plastic debris is a crucial first step toward estimating (and ultimately mitigating) the impact of this insidious and long-lasting pollution on marine ecosystems. This chapter highlights the work of scientists to quantify the spatiotemporal overlap between moving plastic debris and the migration paths of certain species with the goal of modeling and predicting the extent to which different animal species are exposed to moving plastic.

We can surmise through anecdotal evidence that there is a major problem. Quantifying it is another matter altogether. One of the grand challenges associated with understanding the dynamic relationships between marine life and plastic is an acute lack of data on the movement of plastics pollution from the major sources into the so-called *ocean gyres* (the well-known example being the notorious Great Pacific Garbage Patch). Modeling the amount of plastic degradation that happens en route adds another layer of complexity as the original bags, bottles, and countless other items degrade into even more dangerous micro- and nano-plastics.

This work uses geographic information systems (GIS) tools to combine multiple data sources to mine spatiotemporal patterns behind different marine animal species' exposure to moving plastics. The model represents plastic movement as a coupled process between ocean currents and surface winds using a Lagrangian simulator—an open-source algorithm that is widely accepted in the realm of fluid dynamics. Resulting movement of spatially heterogeneous plastics is represented within a space-time data structure. Researchers developed a temporal co-location analysis between plastic movement and animal telemetry to model exposure times of different marine animal species to moving plastics. This type of analysis can be broadly described as *spatial statistics*.

One of the biggest problems with plastics in the oceans is mismanaged plastics traveling into ocean gyres, areas where currents circle and accumulate plastics and other marine debris into so-called garbage patches. Throughout their ocean transport, plastics degrade and form micro-plastics that are detrimental to marine life ranging from the largest whales to the smallest plankton.

PLASTICS IN NUMBERS

8,300,000,000 metric tons of plastics produced since its invention in the 1950s[1]

99.9% of all plastics ever produced still exist in their original shape (including discarded plastics)[2]

Number of microplastics from marine debris in the oceans is **500 times** more than the number of stars in the Milky Way (100 billion)[2]

1.15–2.41 million tons of plastic waste currently enter the ocean every year[3]

74% of emissions occurring between May and October[3]

100s of marine species are at direct risk[2]

Recycled and managed

Unmanaged

Environmentally unfriendly plastics impact the environment when instead of being recycled, they come in contact with the subsystems of our planet. Mismanaged plastic waste is the amount of plastics that cannot be (or are not) recycled with current infrastructure. Rivers, tides, winds, and illegal dumping can all carry mismanaged plastics to the natural environment.

Despite their place at the top of the marine food chain, whales like this breaching humpback are especially vulnerable to ocean plastics as they continually filter sea water and ingest plankton and small fish that in turn may have ingested microplastics.

Up and down the food chain, marine animals, like this crab in the Philippines, face the consequences of plastic products.

MAPPING THE GLOBAL PLASTIC WASTE SITUATION

Data from Jenna Jambeck et al.[4] linked worldwide data on solid waste, population density, and economic status to estimate the mass of land-based plastic waste entering the ocean. The study initially estimated the amount of plastic waste produced per capita. But plastic waste becomes an environmental problem only when it is mismanaged, in other words, when it is not recycled and managed properly.

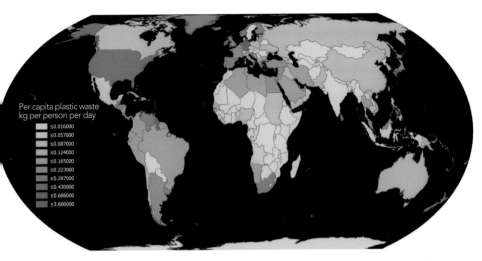

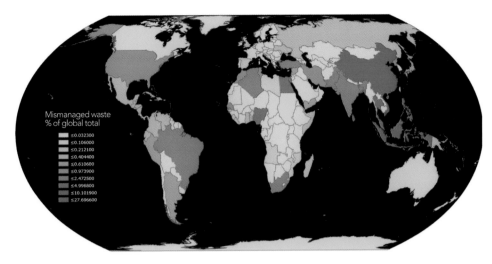

Plastic waste per capita: *This figure displays the amount of plastic waste produced per capita per country. Note that the United States is one of the top producers of plastic waste at an average of 3.6 kg per person per day.*

Percentage of mismanaged plastic per country: *This figure shows that some of the top plastic waste producers are largely managing their plastics waste, preventing it from entering the environment. Amount of plastic waste left in contact with Earth's subsystems is again mapped using the data on the amount of mismanaged plastics by weight.*

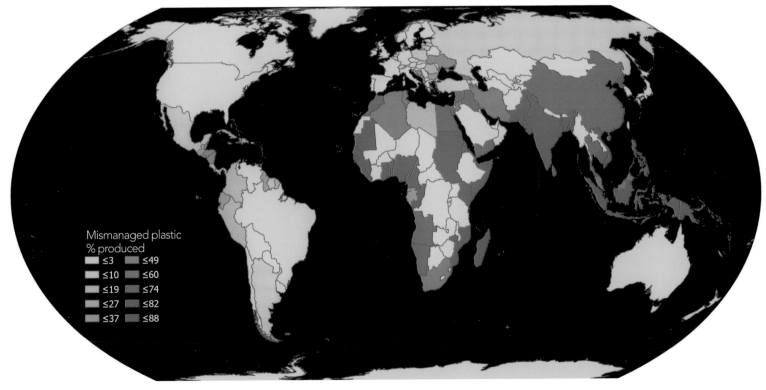

Mismanaged plastic waste: *Countries are symbolized with respect to their percentage contribution to overall plastic pollution. This map displays the contribution of every country to mismanaged plastic waste. Despite the high amount of plastic per capita, North America has a low amount of mismanaged plastic waste. However, other countries with relatively lower plastic waste per capita have higher total contributions of plastic pollution.*

SPATIOTEMPORAL CLUSTERS OF PLASTIC POLLUTION

The already complex variables that come into play when studying this problem are compounded because even *managed* plastic trash doesn't necessarily stay in its country of origin. The maps on the previous page do not reflect transnational flow of plastic waste. Large amounts of plastic waste are being exported to countries in Asia.[5] Thus, the amount of plastic to manage exceeds the amount produced by the local population in Asia.

The first line of defense to protect the environment from plastic pollution is reducing waste by recycling and managing the plastic waste. Managing plastic waste includes these mechanisms:[6]

- Mechanical recycling
- Feedstock recycling
- Incineration with energy recovery
- Landfilling

Mismanaged plastics can pollute the soil[7] and be transported via wind and groundwater into major rivers.[8] Once in a river, plastic debris is transported to an outlet that can put the plastic in contact with the ocean, a sea, or a lake.[9] Once in the ocean, currents typically move the plastic debris along currents until they lose velocity and sink or start converging in areas called gyres.

We can summarize the journey of many plastic molecules from site of use to an ocean gyre in this way:

1. Disposal on land
2. Land runoff into river
3. River runoff into outlet
4. River outlet into ocean
5. Ocean transport to gyre

This quantity of plastic debris is then imported to an open-source Lagrangian simulator—OceanParcels—to model the travel of plastic debris transported by different rivers. The resulting initial data describes movement patterns of plastics, which are then visualized and quantified in ArcGIS Pro.

Pacific gyres

Global positioning system (GPS) tracker data for different marine species provided in the animal telemetry network dataset were filtered to focus on species that travel in the Pacific Ocean. A gyre is characterized as a system of circular ocean current movement (clockwise north of the equator and counterclockwise south of the equator), and with at least 100 GPS points tracking its movement.

Garbage patches are large areas of the ocean where litter, fishing gear, and other debris collect. They are formed by rotating ocean currents called gyres. The Pacific gyre collects garbage between Hawaii and California and deposits this debris into the two major patches seen here. (Graphic courtesy of National Oceanic and Atmospheric Administration.)

MODELING THE JOURNEY OF PLASTIC DEBRIS

The Jambeck study estimated the amount of plastics that rivers carry. It analyzed the movement of simulated plastic particles and animal tracks from river outlets to the Pacific gyres, relying on the Multivariate Clustering tool in ArcGIS Pro. This analysis detects the number of distinct animal movement patterns with respect to the average distance and angle they have to the nearest plastic stream. In addition, analysts used a test of movement correlation based on the existing work by Laurent Lebetron and team[3] that implements the Hidden Markov Model (HMM) in R using an open-source package called moveHMM. The number of distinct movement patterns defined with respect to distance and angle was used in the HMM model to test the significance of impact of plastic streams and gyres to animal movement. This part of the study analyzed whether marine animals were significantly collocated with prominent bodies of moving plastic debris.

Researchers used data for mismanaged plastic from Lebreton[3] to quantify the patterns of plastics arriving at river outlets. The model shows the plastic outputs from 10 rivers in Asia, which contribute a major amount of plastic to the oceans. The datasets were binned in Esri's space-time cube data structure to characterize spatio temporal patterns of plastic pollution originating from these rivers. Following the characterized spatiotemporal patterns, we simulated numerical plastic particles at these eight river outlets with respect to surface winds and geostrophic currents. The simulation was conducted using an open-source simulator that is integrated into ArcGIS Pro through the Python integration via ArcPy. Travel times and routes of plastic particles were analyzed using ArcGIS Pro®.

Because the study covers a global problem, its conclusions refer to that global scale instead of to the plastic pollution management policies of any one country. The analysis shows that 8 of the 10 rivers are positioned geographically where their discharged plastics can reach the Pacific Ocean gyres in less than a decade.

Following the characterized spatiotemporal patterns, the team simulated numerical plastic particles at these eight river outlets and simulates their movement with respect to surface winds and geostrophic currents. The simulation was conducted using an open-source simulator that was integrated into ArcGIS Pro through the Python integration via ArcPy. Travel times and routes of plastic particles were analyzed using ArcGIS Pro.

Last, the team used two methods to analyze the relationships between marine animal paths in the Pacific Ocean and areas of plastic debris concentrations:

- Multivariate clustering in ArcGIS Pro
- Hidden Markov Model in moveHMM package of R

Using the Multivariate Clustering tool, the team detected the number of distinct animal movement patterns with respect to the average distance and angle they have to the nearest plastic stream. The number of distinct movement patterns defined with respect to distance and angle was used in the HMM model to test the significance of impact of plastic streams and gyres to animal movement. This part of the study analyzed whether marine animals were significantly collocated with prominent bodies of moving plastic debris.

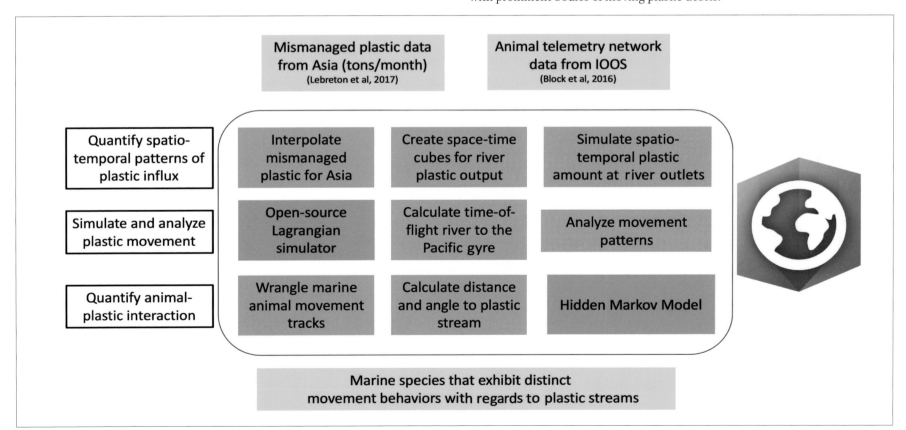

Overall workflow for evaluating the impact of moving plastic debris on marine life. The flexibility of the ArcGIS platform allows incorporating open-source libraries (in orange) in this scientific workflow while enabling the powerful suite of geoprocessing tools (in blue) on various data sources (in yellow). In particular, space-time pattern mining and spatial machine learning tools in ArcGIS Pro are used in the geoprocessing building blocks shown here.

PLASTIC DEBRIS INPUT FROM RIVERS

Plastic debris from rivers is not constant, and seasonality has an impact. Lebreton et al.[3] provided data on inverse-modeled mismanaged plastic by weight globally. Our team uses a subset of this dataset along the coast of Asia, because of its role as an ocean plastic source. Our team creates smooth kriging surfaces of the mismanaged plastic by weight in a time-discrete manner and uses measurements in rivers to estimate the amount of mismanaged plastics flowing through rivers at any given time. Smooth mismanaged plastic by weight surfaces is generated with empirical

Bayesian kriging (EBK).[10] We prefer EBK because it can capture nonstationarity, which is expected because of localizing effects of currents on the plastic debris. The small multiples below depict the resulting smooth surfaces. Notice that they display strong seasonality for plastic waste arriving at the Pacific coast. Note that early in the year, the majority of the plastic pollution originates in the coast of Malaysia and the Philippines and later in the year is surpassed by the plastic outflow from the Yangtze, Pearl, and Yellow rivers.

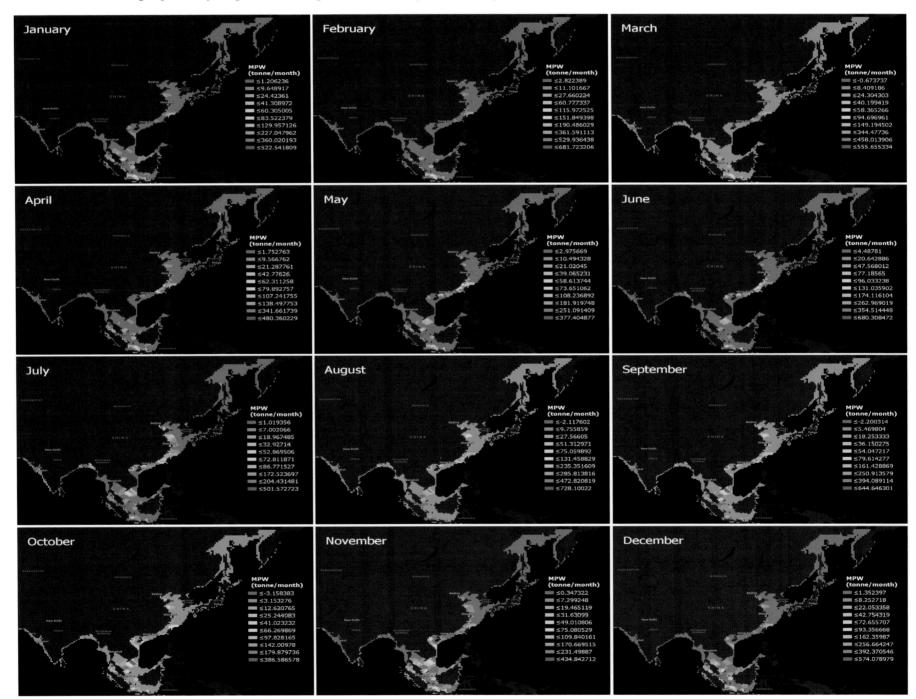

Empirical Bayesian kriging maps of mismanaged plastic by weight per month.

A SPATIOTEMPORAL LOOK AT THE PLASTIC PROBLEM

The Lebretron study[3] estimates that annually between 1.15 and 2.41 million tons of plastic flows from global rivers into the oceans. Their model of plastic inputs from rivers into oceans is based on waste management data, population density maps, and hydrological information. The model is also calibrated against an ever-expanding set of surface plastic field measurements being carried out in response to the recognition of the issue. The top 20 polluting rivers are mostly located in Asia and account for more than two-thirds (67%) of the global annual input while covering 2.2% of the continental surface area and representing 21% of the global population. The data also showed that the majority of plastics is emitted between May and October (roughly correlating with the rainy season).

For the purposes of the GIS part of the analysis, researchers selected the following eight rivers ranked by their estimated mismanaged plastic. The top-polluting rivers in this study set were, in order:

1. *Indus* [India]
2. *Ganges* [India]
3. *Irrawaddy* [Myanmar]
4. *Mekong* [Vietnam, China, Laos, Thailand, Cambodia, Myanmar]
5. *Pearl* [China]
6. *Yangtze* [China]
7. *Yellow* [China]
8. *Amur* [Russia, China]

Mapping in 3D space and time

Researchers aggregated these as space-time bins along the rivers to analyze patterns of incoming pollution to the Pacific Ocean. Resulting hot spot analysis on space-time cubes per river is depicted here.

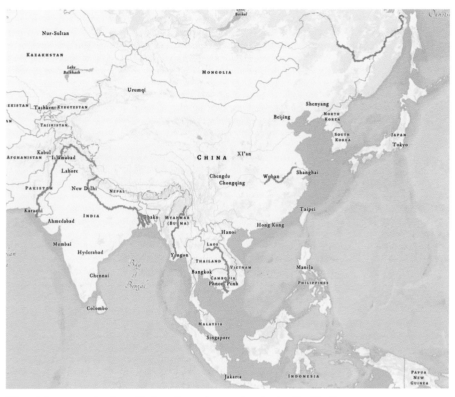

The eight worst plastic-polluting rivers that end up in the Pacific Ocean traverse some of the most densely populated countries on Earth.

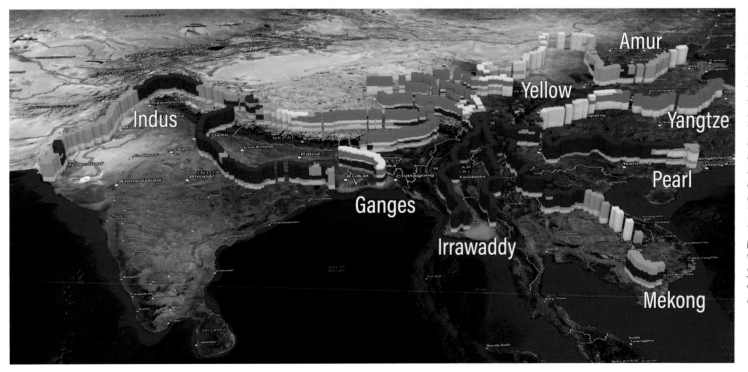

Vertical bins depict time, the lowest bin being January and highest being December. Colors indicate whether the time series was higher than average (red) or lower than average (blue). Emerging hot-spot analysis shows that the Yangtze and Ganges have high plastic debris outflow after May. Note the difference in spatiotemporal patterns of plastic debris mass in the Ganges and Yangtze Rivers. The Ganges exhibits distinct time cycles in which amount of plastic debris is less compared with other rivers earlier in the year. The amount of plastic debris in this river increases later in the year (around May).

HOW FAST DO RIVER-BORNE PLASTICS MOVE?

At this point in the workflow, the focus shifts from the river pollution sources to the global scale ocean surface patterns. Part of the challenge is that closer to shore the patterns are complicated due to the outsized impact of winds on shallow surface waters. The team simulated the movement of plastic particles by using the monthly river outflows of mismanaged plastic as an input to the space-time model. To factor in the impact of all the global patterns, an open-source Lagrangian particle simulator (Open Parcels[11]) was implemented, which simulates the well-known general circulation patterns of Earth's oceans. To account for the impact of wind in shallow currents, another specialized model—HYCOM + NCODA Global 1/12°—was included in the overall model. Particle trajectories are shown here as red traces on the map originating at the outlets of the aforementioned Asian rivers. Note that plastic debris from the Ganges stagnates in the Bay of Bengal.

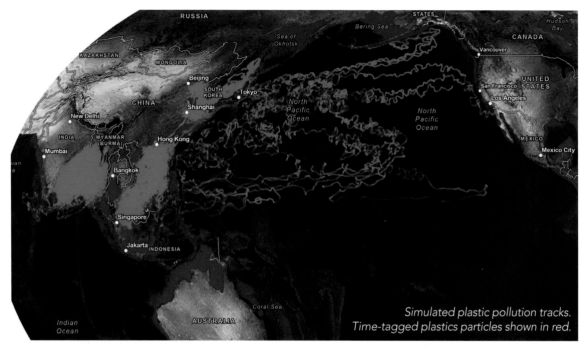

Simulated plastic pollution tracks. Time-tagged plastics particles shown in red.

Another way to examine the data is through spatiotemporal hot spot analysis of the plastic particulate count. Areas of plastic stagnation are depicted as areas of high plastic counts. Note the area west of China that has localized hot spots due to stagnant waters creating high concentration of plastics. In addition, the movement of plastics slows down on certain portions of their trajectories, which is also accounted for in the space-time model. The simulation of plastic particles is terminated when particles reach a coast or a gyre. The map shows new hot spots where plastic streams are accumulating as they move toward gyres.

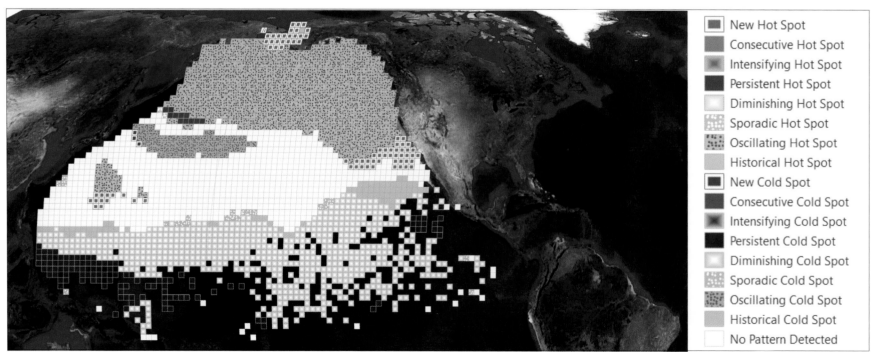

Emerging hot-spot analysis for the Pacific Ocean.

TIME-OF-FLIGHT MAPPING

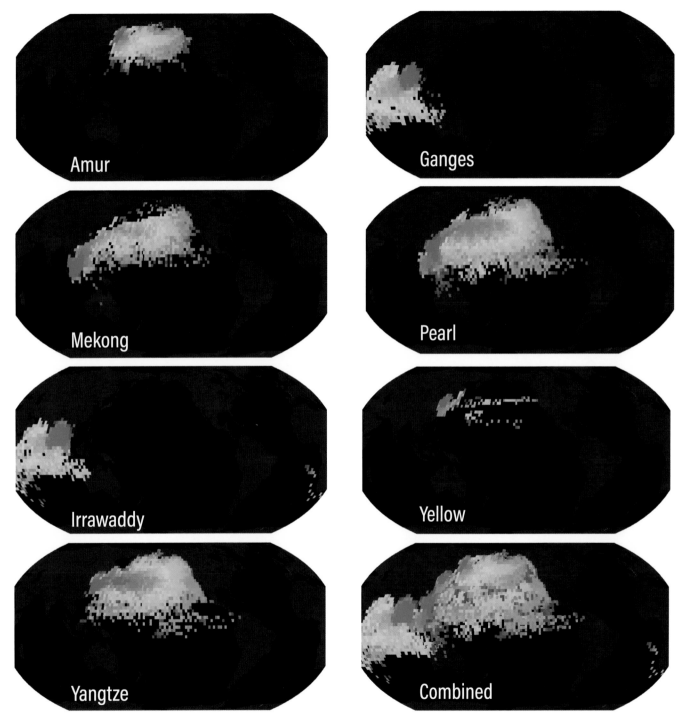

Time-of-flight maps for plastics from different river outlets. Hot colors indicate fast travel times. Hot colors corresponding to low time-of-flights imply plastic emitted at that corresponding river can reach the locations in hot colors faster than cold colors. Note that plastic debris from Ganges stagnates in the Bay of Bengal without reaching the Pacific gyre. From the Yangtze and Mekong Rivers, plastics enter "plastic highways" into the Pacific gyre because islands do not impede the debris as it travels on strong ocean currents. Plastic particles can reach one of the Pacific gyres in two to five years depending on which river it originates from. Thus, reducing the outflow of plastics can save a large water body in the ocean from micro-plastic emitted as these plastic particles travel in the ocean.

MODELING INTERACTIONS BETWEEN PLASTIC MOVEMENT AND MARINE LIFE

Animal telemetry network (ATN) data[10] is the main data source for marine animal movement in the Pacific Ocean. A subset of the ATN network data contains 24 species of marine animals, such as leatherback sea turtles, whales, and marlins, to name a few. Interactions between these marine animals and plastic streams are quantified by calculating the minimum distance between migration paths of marine animals and angle of approach of marine species to plastic streams and gyres.

The team first defined cluster jointly for distance between different species' paths and their angle of approach to the plastic streams. The next figure shows a map resulting from multivariate clustering.

The Calinski-Harabasz index returns four distinct groups in ATN data with respect to distance to plastic stream and angle of movement. Clusters of animal movement are further explored by plotting the characteristics of every cluster with respect to minimum distance to plastic streams and angle. The next figure depicts the box plot for multivariate clusters. For every species, our group also built a Hidden Markov Model to investigate the probability that movement patterns discovered here are statistically significant. R's Hidden Markov Model library is utilized for this purpose.

Clustering map

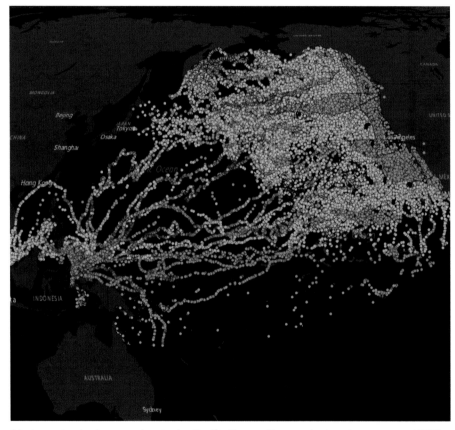

Marine animal telemetry data in the Pacific Ocean. Every point is a time-stamped GPS location for a species. All GPS locations are color coded with respect to clusters defined for distance and angle to plastic streams.

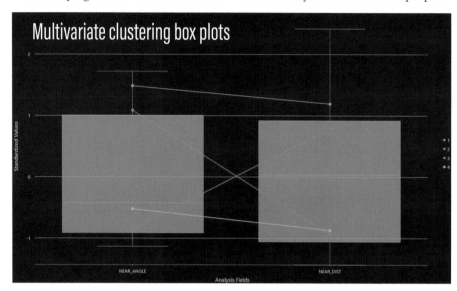

These box-plots show clustered marine animal movement data. Red and yellow indicate species that are collocated with plastic streams that swim with and against the current, respectively. Moving plastic debris is expected to impact these species the most. Blue indicates species that evade plastic currents due to persistent high angle movement and distance. In this instance, the study only investigated the overall angle and distance. Micro-paths may also exist that overall clusters may not reflect. Green indicates species that travel orthogonally to plastic streams. These are plastic-agnostic species that do not alter their path. Most of the species in this group are whales.

Satellite transmitters attached to a sea turtle's back. The transmitters are small, lightweight devices attached to the turtle's carapace (shell) using aquarium-grade epoxy resin and are designed to withstand up to 300 days at sea. The transmitters rely on solar power to charge the unit and satellite telemetry to pinpoint their location every time a turtle returns to the surface for air.

CONCLUSIONS

- Lagrangian simulation results indicate that spatiotemporal contribution of every river to the Pacific plastic gyres cannot be avoided, because time-of-flight from every river varies considerably.

- Preliminary HMM analysis of movement data on marine animals shows that some species alter their course toward plastic streams. Although this analysis does not imply direct correlation simply because these species may travel on the same currents as plastic streams, certain species are observed to be collocated with prominent plastic streams.

- The Yangtze, Mekong, and Pearl Rivers are considered *plastic highways* to Pacific gyres. Plastics emitted from these rivers travel a relatively short distance to reach one of the Pacific gyres.

- Species that manifest distinct movement patterns include the black-footed albatross, northern elephant seal, salmon shark, yellow-fin tuna, and marlin (blue and black). In addition to these species, 20 more species have high collocation with plastic streams, and the angle of their movement is altered around areas of dense plastic.

NOTES

1. R. Geyer, J. R. Jambeck, and K. L. Law, "Production, Use, and Fate of All Plastics Ever Made," *Science Advances* 3, no. 7 (2017): e1700782, https://doi.org/10.1126/sciadv.1700782.

2. United Nations, "'Turn the Tide on Plastic' Urges UN, as Microplastics in the Seas Now Outnumber Stars in Our Galaxy," UN News (2017), accessed January 23, 2019, https://news.un.org/en/story/2017/02/552052-turn-tide-plastic-urges-un-microplastics-seas-now-outnumber-stars-our-galaxy#.WLA81BLyvBJ.

3. L. C. M. Lebreton et al., "River Plastic Emissions to the World's Oceans," *Nature Communications* 8 (2017): 1–10, https://doi.org/10.1038/ncomms15611.

4. J. R. Jambeck et al., "Plastic Waste Inputs from Land into the Ocean," *Science* 347, no. 6223 (2015): 768–71, http://science.sciencemag.org/content/347/6223/768.short.

5. A. Yoshida, "China: The World's Largest Recyclable Waste Importer," *World* 3, no. 4000 (2005): 4–500, http://citeseerx.ist.psu.edu/viewdoc/download?doi=10.1.1.627.2458&rep=rep1&type=pdf.

6. D. Lazarevic et al., "Plastic Waste Management in the Context of a European Recycling Society: Comparing Results and Uncertainties in a Life Cycle Perspective," *Resources, Conservation and Recycling* 55 (2010): 246–59, https://doi.org/10.1016/j.resconrec.2010.09.014.

7. F.M. Li, A.H. Guo, and H. Wei, "Effects of Clear Plastic Film Mulch on Yield of Spring Wheat," *Field Crops Research* 63, no. 1 (1999): 79–86, https://doi.org/10.1016/S0378-4290(99)00027-1.

8. J. M. Coe and D. B. Rogers, *Marine Debris: Sources, Impacts, and Solutions* (New York: Springer, 1997).

9. A. Lechner et al., "The Danube So Colorful: A Potpourri of Plastic Litter Outnumbers Fish Larvae in Europe's Second Largest River," *Environmental Pollution* 188 (2014): 177–81, https://doi.org/10.1016/j.envpol.2014.02.006.

10. K. Krivoruchko, "Empirical Bayesian Kriging Implemented in ArcGIS Geostatistical Analyst" (2012), https://www.esri.com/NEWS/ARCUSER/1012/files/ebk.pdf.

11. M. Lange and E. Van Sebille, "Parcels v0.9: Prototyping a Lagrangian Ocean Analysis Framework for the Petascale Age," *Geoscientific Model Development* 10, no. 11 (2017): 4175–86, https://doi.org/10.5194/gmd-10-4175-2017.

12. B. A. Block et al., "Toward a National Animal Telemetry Network for Aquatic Observations in the United States," *Animal Biotelemetry* 4, no. 1 (2016): 4–11, https://doi.org/10.1186/s40317-015-0092-1.

Plastic Pollution in Ghana photo by Muntaka Chasant.
Humpback whale photo by Whit Welles.

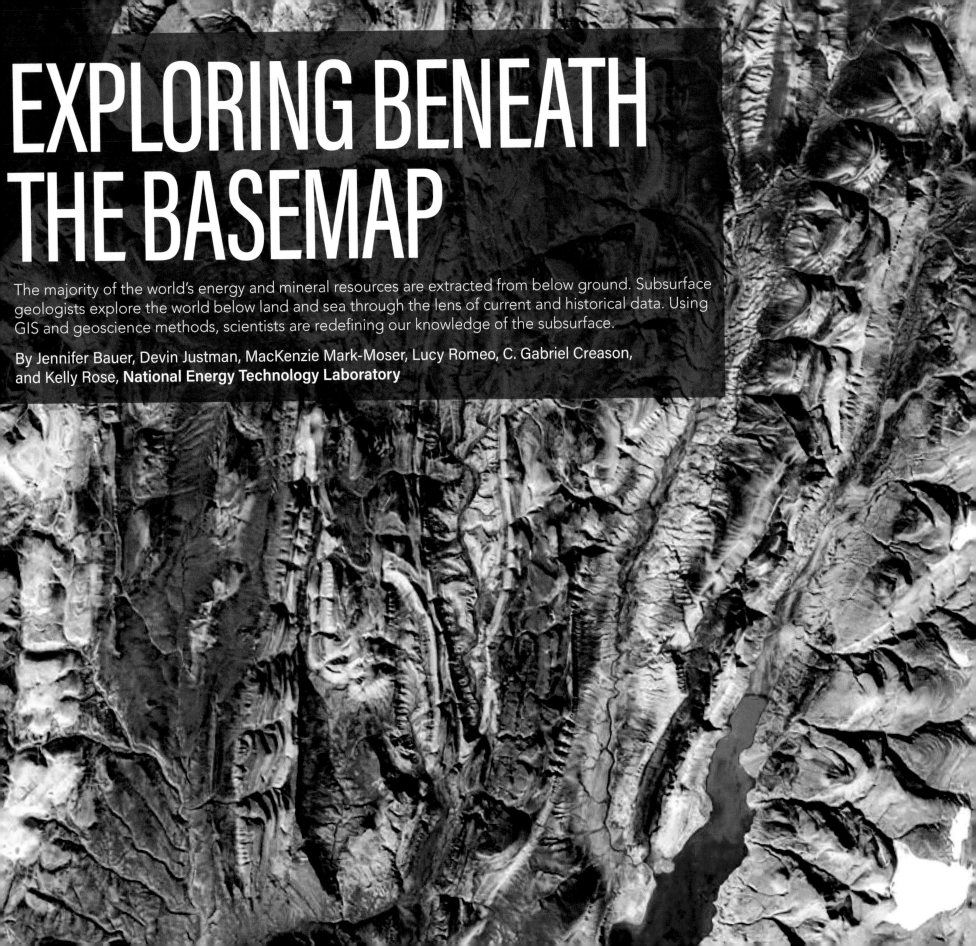

EXPLORING BENEATH THE BASEMAP

The majority of the world's energy and mineral resources are extracted from below ground. Subsurface geologists explore the world below land and sea through the lens of current and historical data. Using GIS and geoscience methods, scientists are redefining our knowledge of the subsurface.

By Jennifer Bauer, Devin Justman, MacKenzie Mark-Moser, Lucy Romeo, C. Gabriel Creason, and Kelly Rose, **National Energy Technology Laboratory**

Layers of exposed rock tell a story about the geology in the Baffin Region, Canada. Subsurface scientists use deep-drilled wellbores and other records to understand the dynamic forces beneath the earth's surface.

EVOLUTION OF SUBSURFACE SCIENCE — FROM GEOLOGY TO GIS

Charles Lyell, one of the founders of the science of geology, is noted for his observation that "the past is the key to the present." Over the past century and a half, subsequent geologists have used the science of cartography to explore, document, analyze, and visualize the geospatial features of Earth's geology. In the past couple decades, scientists have increasingly integrated and utilized GIS tools and techniques to enhance surface exploration and mapping. Efforts to integrate remote sensing data and methods with surface-based observations and measurements resulted in a global basemap of Earth's geology.

However, extending traditional GIS map analyses and interpretation methods into the subsurface is challenging. The geology of Earth is heterogeneous, and the geologic features observed today are the result of complex, systematic processes that have occurred over the past thousands to billions of years. Much of our understanding of the subsurface relies on limited physical samples or from indirect measurements, such as geophysical surveys and wellbore logs. However, data collected with indirect measurements are uncertain, and the accuracy of the measurements declines when used in the deeper regions of the subsurface, especially for the earth's mantle and core.

Analytics, such as those applied to predict, interpolate, and map subsurface properties, such as temperature, pressure, porosity, permeability, and others, rely upon spatially and temporally disparate data coupled with limited a priori information. Resulting predictions are often highly variable, with poorly constrained values and high degrees of uncertainty. Even regions with concentrated subsurface exploration are still plagued with geologic uncertainty that can obstruct safe and efficient exploration of the subsurface. As a result, even with an increase in human exploration of the subsurface, the earth's vast interior remains largely unexplored. There is, however, great potential for GIS-geoscience innovations to address this uncertainty and improve our exploration and study of the earth's subsurface to meet a variety of challenges.

By far, human exploration of the subsurface for mining, water, oil, natural gas, geothermal energy, underground storage, and research purposes has provided most of the opportunities to obtain the direct measurements, information, and samples that have allowed researchers to create a more complete understanding of Earth's crust. The scale of human interactions into Earth's crust varies across the globe, both spatially and temporally. Hard rock mining and wells drilled for drinking water and geotechnical purposes account for centuries of relatively shallow subsurface interactions, penetrating only the upper hundreds of meters of Earth's crust. However, the millions of wells drilled for energy exploration for oil, gas, and geothermal

resources, as well as for underground injection and research across the world, offer greater insight into deeper portions of Earth's crust. Some of these wells reach depths greater than 40,000 ft (or more than 12,000 m) below the surface, but most, on average, are within 10,000 to 20,000 ft (3,000 to 6,000 m) range. All these conduits into the subsurface offer opportunities to collect data that can be utilized to improve characterization and mapping of the Earth's subsurface systems.

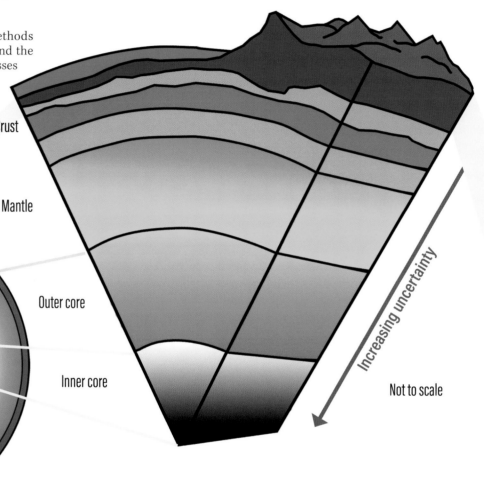

Crust

Mantle

Outer core

Inner core

Increasing uncertainty

Not to scale

The exploration and study of the Earth's surface have provided key insights into the geology of the planet below the basemap. However, extrapolation and prediction of subsurface features are based on limited spatially and temporally dispersed surface observations and indirect measurements. These data are limited by uncertainty, which increases the further from the surface that humans delve.

Data to describe subsurface geologic features come from numerous sources. Surface-based observations as well as samples (i.e., fluid and rock) brought to the surface from subsurface mining and coring efforts offer insights about subsurface systems and processes. These types of direct measurements are often used to cross-validate interpretations and analyses performed using other subsurface data types. Surface-based geophysical studies provide indirect measurements of subsurface characteristics and properties, which are often used to predict properties and patterns in Earth's subsurface systems. Wells serve as the primary source of finer-scale data, offering direct, high-resolution measurements of in situ subsurface properties in Earth's crust. Data from wells fed numerous geologic, geostatistical, and GIS-based studies to improve prediction and constrain our understanding of the in situ geology of Earth's crust and how human interactions have perturbed these subsurface systems over time.

The challenges to explore, analyze, and visualize the subsurface with current techniques have resulted in a demand for better data- and knowledge-driven methods to improve the prediction of subsurface properties. GIS offers a solution to integrate indirect and direct measurements to better constrain the subsurface architecture and evaluate the distribution of subsurface resources. GIS and geostatistical methods, when coupled with data from direct and indirect measurement, offer solutions to further improve our ability to predict, explore, and evaluate the subsurface.

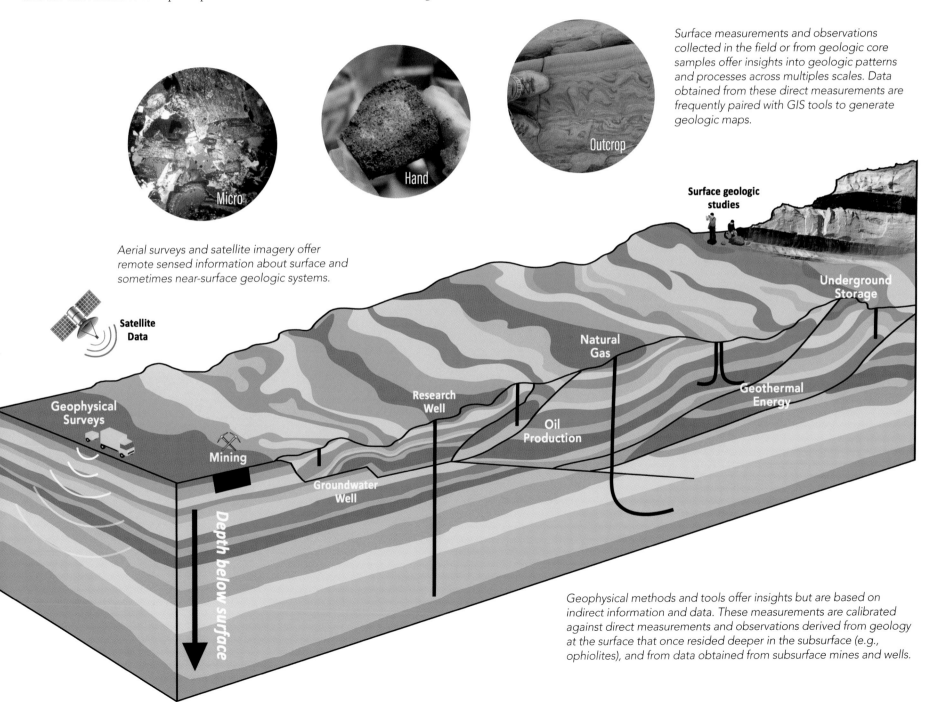

Surface measurements and observations collected in the field or from geologic core samples offer insights into geologic patterns and processes across multiples scales. Data obtained from these direct measurements are frequently paired with GIS tools to generate geologic maps.

Micro

Hand

Outcrop

Surface geologic studies

Aerial surveys and satellite imagery offer remote sensed information about surface and sometimes near-surface geologic systems.

Underground Storage

Satellite Data

Natural Gas

Geothermal Energy

Geophysical Surveys

Research Well

Oil Production

Mining

Depth below surface

Groundwater Well

Geophysical methods and tools offer insights but are based on indirect information and data. These measurements are calibrated against direct measurements and observations derived from geology at the surface that once resided deeper in the subsurface (e.g., ophiolites), and from data obtained from subsurface mines and wells.

HISTORY OF SUBSURFACE EXPLORATION

Subsurface drilling for scientific exploration, resource production, and waste disposal worldwide has existed since the 19th century. Deep subsurface drilling activities began in the mid-1800s, targeting the commercial production of salt brines in the present-day countries of Canada, China, Poland, and the United States.[1] In 1859, the first commercial well targeting oil, the Drake well, was drilled near Oil Creek, Pennsylvania.[1,2] The realization that mineral oil had commercial value helped drive the persistent expansion of the deep subsurface drilling record. This record represents one of GIS and geoscience's best foundational datasets for investigating and interrogating subsurface systems. To date, there are more than 6 million deep drilling well records worldwide associated with oil, gas, geothermal, underground fluid disposal, research, and permitted locations. These wells are found across all continents, except Antarctica, and scattered across the world's oceans. More than half of these wells penetrate more than 3,000 ft (1,000 m) into the subsurface.[3]

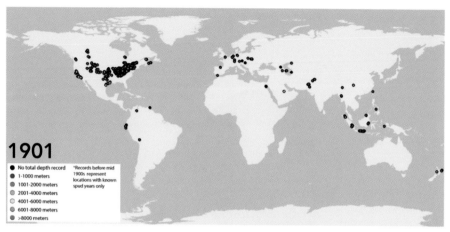

1859

- No total depth record
- 1-1000 meters
- 1001-2000 meters
- 2001-4000 meters
- 4001-6000 meters
- 6001-8000 meters
- >8000 meters

*Records before mid 1900s represent locations with known spud years only

In the mid-1800s, the spatial footprint of deep subsurface exploration begins to emerge, consisting of deep wells drilled for salt brine production in North America, Europe, and Asia. Deep well drilling for commercial oil production also begins. Drake well denoted by the red star on the map above.

1901

- No total depth record
- 1-1000 meters
- 1001-2000 meters
- 2001-4000 meters
- 4001-6000 meters
- 6001-8000 meters
- >8000 meters

*Records before mid 1900s represent locations with known spud years only

By the early 1900s, deep drilling mainly becomes tied to oil exploration, resulting in an increase in subsurface activities. The footprint of drilling expands across North America, South America, Europe, Asia, and parts of Oceana, with visible spatial clusters beginning to form in areas where oil prospects are rich across the United States.

| 1850 | 1860 | 1870 | 1880 | 1890 | 1900 | 1910 | 1920 | 1930 |

Retouched 1859 photograph showing Edwin L. Drake, right, and his Drake Well, the first commercial oil well drilled in the United States.

In 1901, with the automobile era under way, the new "black gold" was discovered at Spindletop in Texas. Within a year, hundreds of wells were crowded together, one next to another, as prospectors rushed to discover the next gusher.

The lack of robust data, information, and mapping-based methods to explore the subsurface is evident from this pattern of drilling in that era, as seen in the image on the left from Spindletop, Texas. Most prospectors were drilling wells, sometimes within 10 feet of an existing well, reflecting the lack of systematic, data-driven mapping and predictions that are the hallmark of today's subsurface geo-discovery efforts (right).

Systematic, data-driven assessments spur change

From 1960 to 2010, the chance of drilling a "dry hole," or a well that did not contain oil or gas, dropped from more than 40% to about 10%. This improved efficiency is directly related to the use of new and improved technologies, such as seismic surveys, subsurface imaging, and improved interpolation methods, such as geostatistics, which help improve the prediction of subsurface properties. New technology coupled with systematic, data-driven methods that integrate GIS help improve subsurface characterization and more accurately detect patterns and trends in the subsurface, helping transform the way geoscientists assess and map Earth's subsurface. By the 1960s, innovative technologies began to increase the footprint of drilling across the globe, touching every continent except Antarctica. Modern activities have expanded the cumulative footprint of subsurface drilling to all the continents and oceans.

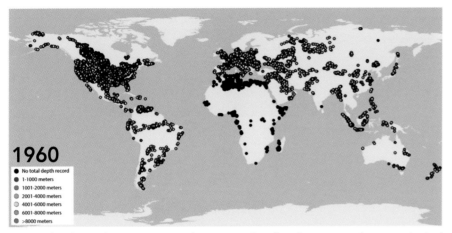

New technologies for measuring and imaging subsurface features, such as geophysical surveys, coring, and geophysical logging of wellbores, increased in use during the 1960s, and demonstrated how activities could benefit from more systematic efforts to characterize, predict, and map subsurface features.

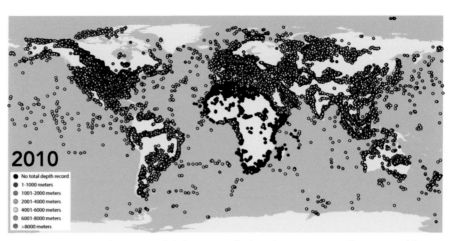

From 1960 to 2010, the successful discovery of oil and gas in new wells increased from ~40% to ~90%.[4] The increase of subsurface research and characterization and the integration of geologic exploration with geophysics and GIS methods have helped significantly improve our understanding of the subsurface.

1940   1950   1960   1970   1980   1990   2000   2010   2020

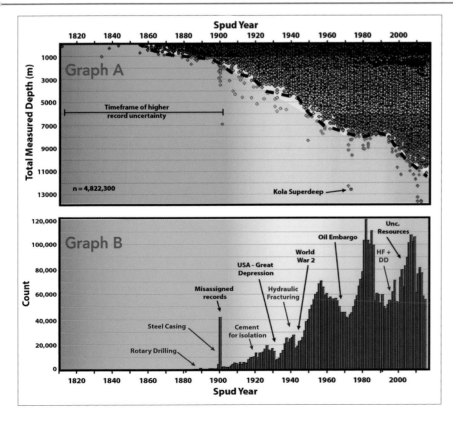

Six million wells were recorded between 1802 to 2015 (footprint displayed on the map).[3] Through these records, wells were drilled at increasing depths (graph A) over time, based on the reported "spud" year (or the year the drill bit begins drilling a well). The number of wells drilled each year fluctuated greatly over time (graph B), showing the "boom and bust" nature of these activities and influence of technology advances (in blue) and major historical events (in black).[3] This entire historical record offers a trove of data for modern geoscientists, providing critical insights to better support geo-discovery, geo-exploration, and geo-hazard prevention when coupled with geoscience and GIS methods.

GEO-DISCOVERY: AN ENDURING RECORD IN THE EARTH'S DEEP CRUST

The drilling of more than 6 million deep wells during the past two centuries has left an enduring global footprint of human alteration of the subsurface. These activities have resulted in the placement of wells on all major oceans and continents, reaching down more than 40,000 ft (or 12,000 m) into the Earth's crust.

Beyond simple exploration and characterization of the planet's subsurface, human engineering of the subsurface has resulted in far-reaching and indelible changes to the in situ geology of Earth's crust. Carbon storage, geothermal resource production, underground fluid injection, compressed air storage, drinking water production, agricultural water storage, and natural gas storage encompass a growing suite of activities taking place in the subsurface worldwide.

Ultimately, this magnitude of subsurface exploration and engineering has changed the composition and the behavior of the subsurface itself. The hybridization of geologic, geophysical, geostatistical, and GIS methods offers a unique solution to explore beneath the basemap, providing new insights for *geo-discovery* and ways to improve analysis and exploration of our planet's *geo-hazards* and *geo-resources*.

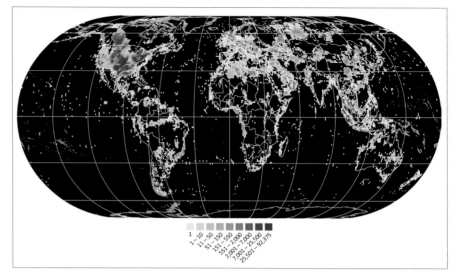

Total density of wells per 1-degree cell as of 2015.

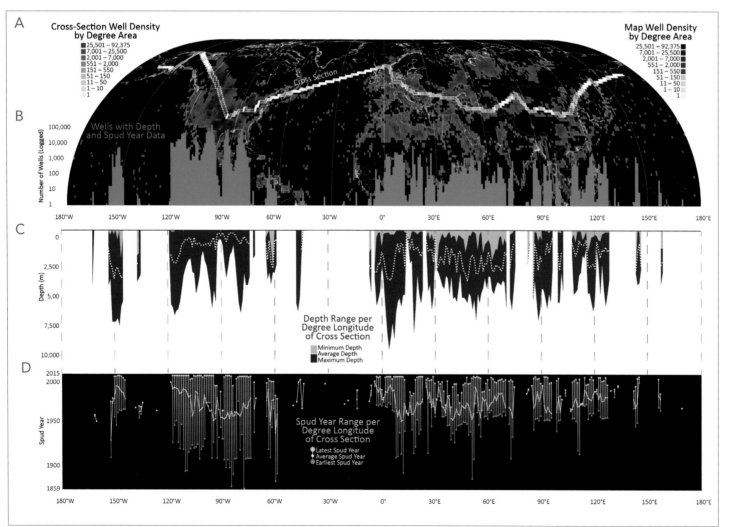

Global cross section highlighting the range and variations in well density, depth, and age across the globe. (A) Well density per 1-degree grid cell, with a highlighted line that is used to create a cross section through the Northern Hemisphere for (B) the corresponding histogram of total number of wells, (C) to plot the minimum, maximum, and average drilling depth for wells in each 1-degree area along the cross-section line, and (D) to plot showing the minimum, maximum, and average spud year (or year that the drill bit begins to drill a new well) for wells in each 1-degree area along the highlighted cross-section line.

Subsurface data boom

The rapid worldwide expansion in subsurface exploration coupled with innovative technologies to obtain subsurface signals, remote sensing tools, and GPS-enabled devices and equipment have also spurred an exponential boom in subsurface data.

This deluge of subsurface data has produced extensive volumes and varieties of high-resolution and quality subsurface signals, measurements, observations, and modeled data. However, spatial and temporal coverage and overlap of these data are highly varied, often forming a patchwork of different data types, formats, resolutions, ages, and quality within the area of interest. The heterogeneity of subsurface data presents a challenge in determining the best process to use to synthesize these large volumes and varieties of data to drive novel analyses.

To effectively and efficiently filter through the subsurface data deluge, scientists at the Department of Energy's (DOE) National Energy Technology Laboratory (NETL)

are using data-science computing methods paired with GIS and other open-source data-processing tools, custom scripts, and machine learning to tease out pertinent information to support interpretation and analytics for subsurface data.

This rapid growth and availability of subsurface data, resulting from increased human exploration, offer an opportunity to produce better data- and knowledge-driven methods with the integration of GIS. Spatiotemporal statistics, geostatistics, machine learning, and artificial intelligence (AI) offer prospective techniques that improve prediction of the subsurface beyond the limitations of current methods, which rely on disparate and limited a priori information. In the geo-resources and geo-hazards sections of this chapter, the intersection between geostatistics, geology, and GIS are further examined with examples of how these three disciplines combine to improve forecasting and insights into subsurface systems.

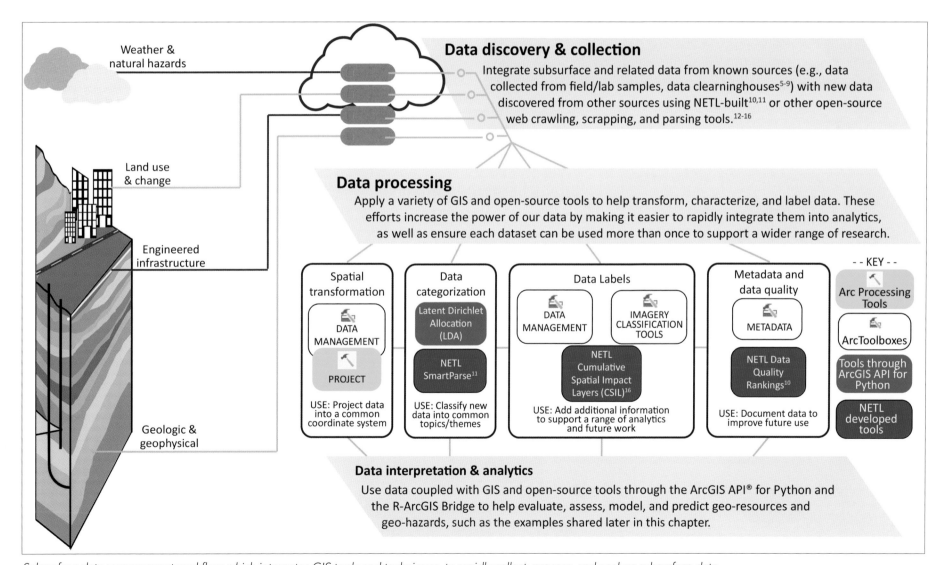

Subsurface data management workflow, which integrates GIS tools and techniques, to rapidly collect, process, and analyze subsurface data.

GEO-DISCOVERY: GIS AND GEOSCIENCE FOR SUBSURFACE PROPERTY PREDICTION

One of the most infamous geologic hazard events, the Deepwater Horizon oil spill, was initiated by a catastrophic wellbore blowout in April 2010.

Traditional 2D surface geology maps have long been coupled with cross-sectional figures to help visualize geologic data beneath Earth's surface. Now, through the power of geostatistics and GIS, we can combine data from direct surface-based geologic observations with subsurface data collected using indirect methods, including seismic data and wellbore geophysical measurements. Coupling geostatistics and GIS has expanded our ability to visualize in three dimensions and improve prediction of subsurface properties, such as temperature, pressure, porosity, and others. This ability is critical to evaluating geologic resources, such as reservoir sands and geothermal plays, and geo-hazards, such as wellbore blowouts and shallow gas accumulation, all of which have implications for ensuring the safe and responsible use of natural resources in the future.

Integrating geologic information and geostatistics is most effective when approached methodically.[17] Without geologic context, statistical methods fail to accurately predict subsurface properties, largely because of the heterogeneous and highly variable subsurface environment. The Subsurface Trend Analysis (STA) framework was developed to pair geologic information, geostatistics, and GIS together to improve subsurface property prediction. By integrating geologic history and context into our predictions, we receive better predictions over larger areas for the highly variable subsurface. Our use-case in the Gulf of Mexico demonstrates how we applied the STA to improve interpolations for reservoir sand pressure and temperature gradients.[17,18]

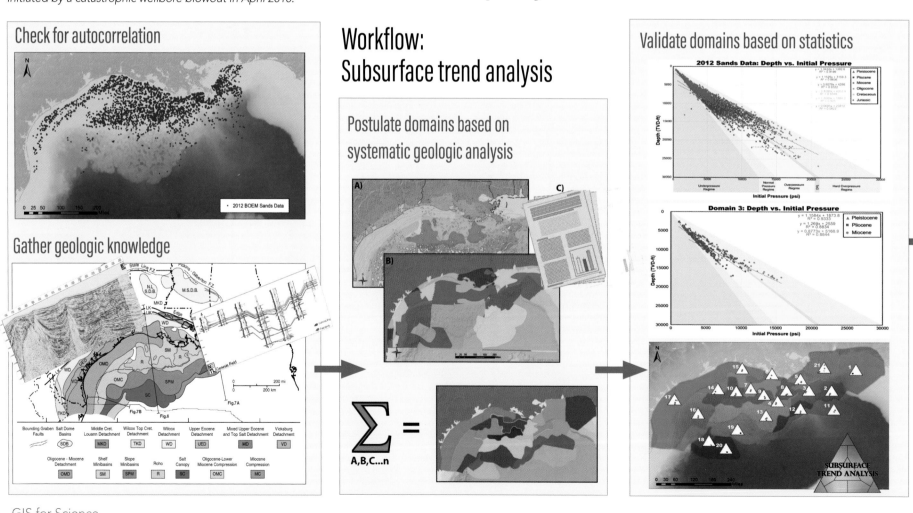

Advanced property prediction and feature analysis

The initial use-case of the STA method improved prediction of the subsurface pressure gradient for oil and gas reservoir sands (A). Predicted values were cross-validated with 150 new data points, and results showed that the STA improved the prediction of subsurface pressure gradient for two out of every three new data points when compared to predicted pressure gradients interpolated using empirical Bayesian kriging (EBK).[17] In addition, we were able to use STA to predict additional subsurface properties, such as temperature gradient (B), and compare interpolation trends with the presence of certain geologic features, such as natural seeps that allow oil and gas to migrate up to the seafloor. Improved prediction of subsurface properties and the presence of geologic features that can result in human and environmental hazards offer critical insight to ensure safe and efficient energy production and geologic storage, as well as supporting better decision-making to help protect the environment and energy economy.

The subsurface seeps that support unique seafloor ecosystems, such as the chemosynthetic community studied by the Alvin deep sea submersible shown at left, can indicate subsurface conditions that contribute to geologic hazards.

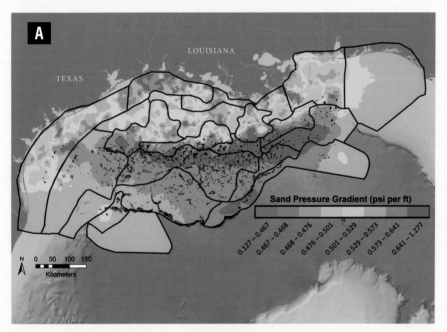

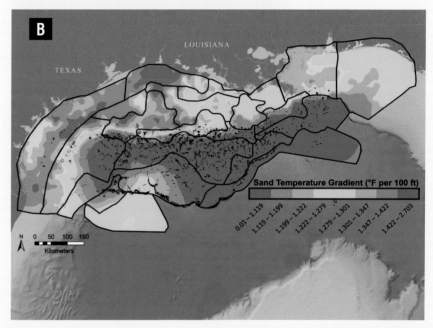

Predicted reservoir sand pressure gradient (A) and sand temperature gradient (B) in the offshore Gulf of Mexico performed with influence from Subsurface Trend Analysis domains (polygons), compared to the location of known and suspected subsurface seeps (points).

Multienvironment, multiscale prediction

The STA framework has been applied to provide subsurface property predictions beyond oil, gas, and geologic storage applications. One such effort involved integrating the STA approach into an assessment method that evaluates and predicts rare-earth element accumulations in sedimentary lithofacies (the REE-SED Assessment Method). The STA approach enables assignment of properties for defined geologic domains, in turn reducing geologic uncertainty. This improved understanding of the spatial distribution and concentration of REE's offers critical insights to ensure effective extraction from discarded coal tailings and sedimentary systems in the future.[19] Utilizing the full spectrum of subsurface information afforded by the STA approach—from direct and indirect data to contextual information from literature—can support sustainable subsurface resource use for decades to come.

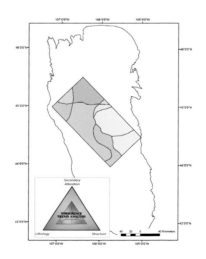

The STA approach utilized as a step in the REE-SED Assessment Method. Example shown here where distinct STA domains (colored areas) are defined for a region in the Powder River Basin, Wyoming (black outline).

GEO-HAZARDS: IDENTIFYING AND MAPPING RISK

Subsurface geo-hazards include earthquakes, faults, fractures, and seeps of gas and other fluids. These geo-hazards are natural geologic processes that can occur at multiple spatial and temporal scales. Each poses a unique risk to the environment, people, and property. Human interactions with the subsurface often perturb the state of these geologic systems and processes. Understanding human interactions in the subsurface and how they affect current geologic conditions can help characterize geo-hazard risk. Integrating big data volumes of different geologic, geophysical, and subsurface properties with GIS tools provides insights that can help mitigate, prevent, and prepare for future hazards.

Induced seismicity

Cases of induced seismicity, or earthquakes that are caused by human-related activities, have been documented around the world. Several cases have been linked to different subsurface activities, such as waste-water disposal, hydraulic fracturing, oil and gas production, dams (reservoir impoundment), geothermal operations, and mining. A study of 198 induced earthquake cases that have occurred since 1929 suggests that these practices have potentially caused earthquakes with magnitudes as high as 7.9.[20] Since 2008, Oklahoma has seen a dramatic increase in the number of earthquakes with magnitudes ranging from 3 to greater than 5, with many occurring on previously unknown and unmapped faults. These events are mostly attributed to the increase in wastewater disposal volumes, a byproduct of oil and gas operations within the state. Research suggests that the recent spike in earthquakes in Oklahoma can be linked to a range of human interactions with the subsurface, and other natural factors.[21]

Efforts to characterize geo-hazards, such as induced seismicity, can benefit from an understanding of key patterns and trends between faults, earthquakes, and other subsurface properties. Coupling geospatial analytics with big volumes of subsurface data offers a novel workflow to predict subsurface faults and fractures, especially for areas with little or no data, and help improve our understanding of subsurface geo-hazards.

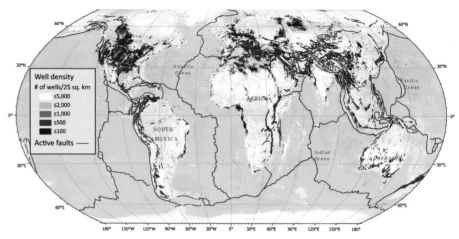

Global distribution of wells[22] and active faults that may become a source for an earthquake.[23]

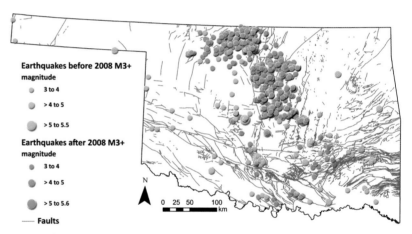

Location of earthquakes[24] before 2008 (orange) and after 2008 (red) in relation to subsurface faults[25] (green lines) in Oklahoma.

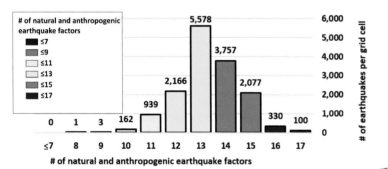

Number of earthquakes (after 2010; scaled by 3D height of grid cell) per number of natural and anthropogenic earthquake factors within 100 sq. mi. grid cells in Oklahoma.

Workflow: Predicting faults and fractures

Efforts to predict subsurface faults and fractures, or geologic structural complexity,[26] aim to offer improved geo-hazard prevention. But these efforts are limited by ambiguous and sparse data that often lack key attribute information. To overcome these challenges, a coupled data-driven, fuzzy-logic, and GIS workflow was developed to predict subsurface structural complexity (SC).[27] The approach helps to improve the characterization of the subsurface and offers critical information for predicting and prevention of geo-hazards, especially in areas where little or no data on subsurface faults and fractures exist.

To determine structural complexity, explicit fault and earthquake location data were used as labels of known structural complexity in the fuzzy logic model (steps 1 and 2 in the workflow). These data were then used to train and test the model (steps 3 and 5) with other topographic, lithologic, and geophysical proxy datasets (step 4) to predict structural complexity. The predictions (generated from step 6) were evaluated (steps 7 and 8). Results demonstrate the model's effectiveness and limitations as a screening approach to identify and validate structurally complex areas in maps and interpreted cross sections.

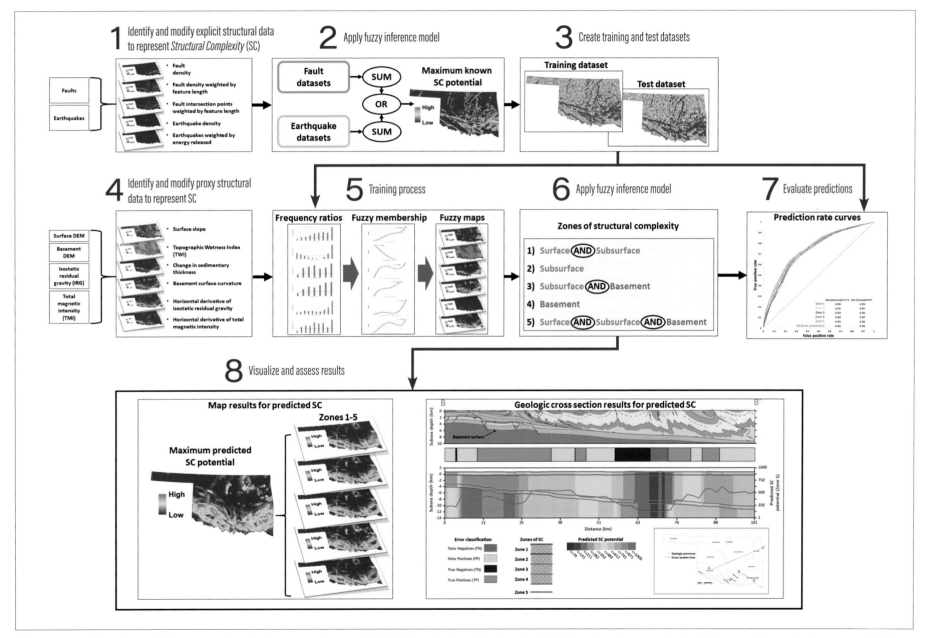

Illustrative workflow for predicting areas of geologic structural complexity (SC; i.e., faults and fractures) within the subsurface, applied in Oklahoma. The geologic cross section[28] (bottom right) is aligned to predicted SC for each zone (green to red scale). Model results are compared and shown as True Positives (correct prediction of SC; light blue), True Negatives (correct prediction of no SC; dark blue), False Positives (incorrect prediction of SC; pink), and False Negatives (unknown accuracy of SC prediction due to data limitations; red) along the cross section as horizontal bars.

GEO-RESOURCES: RETHINKING SUBSURFACE RESOURCE EXPLORATION FOR STORAGE

With new technology, depleted oil and gas reservoirs that were initially mapped and analyzed for oil and gas production can be reused for a variety of storage needs, including geothermal, water, and carbon capture and storage (CCS). In relation to greenhouse gases, such as carbon dioxide (CO_2), the fossil fuel and industrial processes make up ~65% of global emissions,[29] and efforts to reuse depleted subsurface reservoirs for storage, especially for CCS, offer a solution to help mitigate emissions. The use of depleted reservoirs for CSS requires the presence of a geologic seal to trap CO_2 once injected, the absence of leakage pathways to limit risk, existing infrastructure for transport and injection, and for the area to be able to meet an economically efficient storage capacity. Uncertainty and risk are inherent with CSS, including concerns over groundwater contamination and leakage of CO_2 up to populated areas, which can have significant human and ecological impacts. Evaluating the potential for successful CCS sites relies on data, as it is a data-driven process, where the assessment of each project will only be as good as the data used to assess the project site.

> 20,000	
5,000 – 19,	
500 – 4,99	
50 – 499	
10 – 49	
5 – 9	
1	

Density of potential sources of CO_2 emissions, which includes oil and gas infrastructure (wells platforms and well pads, refineries, processing plants, power plants, liquefied natural gas (LNG) terminals, pipelines, mines, stations and storage sites), oil, gas, electrical, industrial and public infrastructure, agriculture processing sites, ethanol and cement manufacturing sites, fertilizer and ammonia production plants, and waste management sites.[10,20,31]

Differences in the subsurface: Onshore versus offshore

Storage capacity has been characterized and predicted for multiple regions, which includes understanding where storage could occur and how this solution to greenhouse gas mitigation differs onshore versus offshore. The National Risk Assessment Partnership[30] (NRAP) applies science-based predictions in engineered-natural systems to assess the safe and long-term storage potential of CSS. This partnership is composed of members from five national labs including NETL, Lawrence Berkeley National Laboratory, Lawrence Livermore National Laboratory, Los Alamos National Laboratory, and the Pacific Northwest National Laboratory. With input from industry, academia, governmental, and nongovernmental organizations, this initiative has built the resources needed to understand how storage systems behave in the extreme subsurface conditions over time.

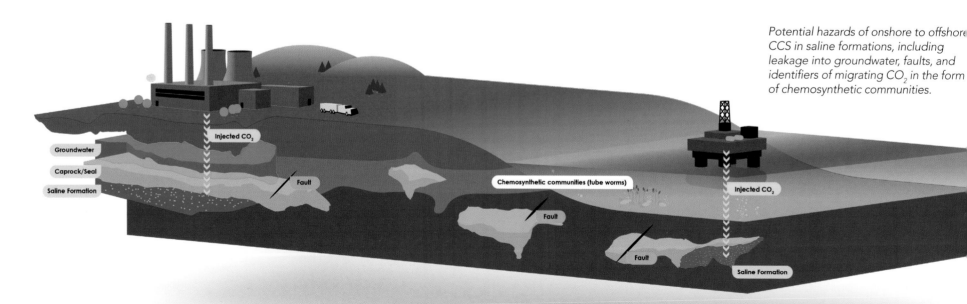

Potential hazards of onshore to offshore CCS in saline formations, including leakage into groundwater, faults, and identifiers of migrating CO_2 in the form of chemosynthetic communities.

Offshore CSS, which has been explored but is yet to be proven in US waters, is farther from human population centers and groundwater sources, but still presents risks because of increased data uncertainty and leakage potential to effect surrounding environments. The offshore environment for storage differs from the onshore with frequently changing pressure-temperature regimes, lithologies (i.e., type of rock), and depositional settings at multiple scales throughout the 3D space.

For example, the Gulf of Mexico subsurface is more unlithified (i.e., less solid, more porous) and unconsolidated than onshore areas. Taking these differences into consideration, NETL has implemented the DOE volumetric methodology to calculate storage capacity potential for geologically distinct domains in the offshore Gulf of Mexico. These domains are spatially defined and statistically supported by NETL's STA method. These efforts have built up spatial data resources for better capacity prediction and implemented GIS techniques to select injection sites and areas, avoiding known leakage pathways, identifying applicable infrastructure, and calculating storage resource potential.

Additional opportunities for reservoir reuse

In addition to CCS, other strategic uses for the reservoirs include storage for fluid waste or compressed air, which essentially operates as subsurface battery, and enhanced oil recovery. Enhanced oil recovery is the process of injecting materials into existing wells to increase pressure on the trapped oil, lowering the viscosity, making it easier to recover. With increasing amounts of accurate data, advanced spatial analytics, and an understanding of geologic processes, oil and gas reservoirs can be sustainably repurposed for CCS and other resources with less uncertainty and risk.

Sediment Thickness (km)
>16
14
12
10
8   Potential Storage Site
6   Potential Capture & Storage Site
4   Sedimentary Basins
2
0

Identified potential sites for CCS in sedimentary basins, which include capture and storage in basalt formations, oil and gas reservoirs, shales, saline formations, unminable coal areas, and others. Sites are laid over global onshore and offshore sedimentary thicknesses ranging from 0 to 18 km.[32,33]

GEO-RESOURCES: RETHINKING COAL RESOURCES TO DRIVE ENVIRONMENTAL AND ECONOMIC BENEFITS

Coal source footprint

A spatiotemporal understanding of the life cycle of coal resources is essential to optimize their use and inform resource management. Integrating geospatial data related to coal production, delivery, consumption, and waste streams allows us to characterize and evaluate coal as a resource. Coal has been extracted from thousands of mines throughout the United States that lie within defined regions or fields that have experienced similar geologic histories. Moreover, coals within these regions often display similar physical and chemical attributes that may be optimal for specific uses, including electric power generation, heating, steel manufacturing, carbon-based products, and other industrial processes.

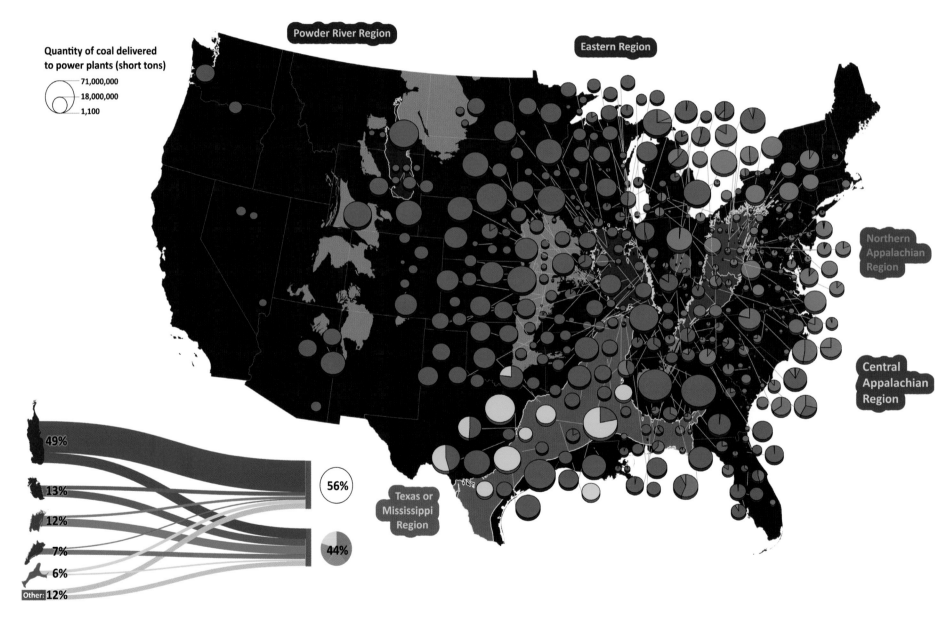

Quantity of coal delivered from the top five source regions[34] to power plants between 2011 and 2016.[35] Both figures are colored by source region, and the size of the pie chart (US map) or line thickness (Sankey diagram; left) represents the quantity of coal delivered. In the map, the color of each pie chart represents the percentage of coal received from each region. In the Sankey diagram, percentages denote the total percentage of coal received at power plants from each region on the left to the percentage of power plants that receive coal from a single region (uncolored circle; top right) versus multiple regions (colored circle; bottom right) on the right.

Matching coals to associated industry needs requires transportation from the mine to the associated facility. From 2011 through 2016, a total of 4.8 billion short tons of coal was delivered to plants for electric power generation throughout the United States. Approximately half (49%) of this coal was delivered from the Powder River Region, an area straddling the border of Wyoming and Montana. Depending on the requirements of coal attributes needed for a specific use, a power plant or facility may source its coal from one or more regions.

During the same time period, slightly more than half (56%) of all coal delivered was received by power plants that source coal from a single region, accounting for 59% of all power plants, and while 46% of delivered coal was received by power plants that source coal from multiple regions, accounting for 41% of all power plants. The mixing of coals at individual facilities and industries can inform and optimize coal delivery networks from mine to facilities and supports understanding of coal byproducts or postcombustion waste streams as a resource.

Postcombustion waste streams

After coal is burned at a power plant or facility, significant volumes of waste streams or byproducts are produced, being one of the largest sources of industrial waste in the United States. Of the approximately 754 million tons of byproduct produced from 2011 through 2016, 67% was coal ash and the remaining consisted of materials such as gypsum, boiler slag, and others. Most coal ash was disposed of in landfills, ponds, and offsite locations (58%) or sold for beneficial use (38%), mainly in concrete and cement products, mining applications, or as structural fill.

Understanding the fate and use of coal and coal byproducts supports opportunities to increase beneficial use or remediate disposal locations of coal ash. These opportunities can provide potential economic benefits, reduce environmental hazards, and optimize coal as a resource.

Map displays power plants[35] as pie charts colored by percentage of coal ash disposition and sized by coal ash quantity produced from 2011 to 2016.[36] Disposition is categorized by disposal (red), beneficial use (green), and storage (blue). Leader lines point to location of power plants (gray lines). The cumulative breakdown of disposition for the contiguous United States between 2011 to 2016 is represented by the larger pie chart on the bottom left.

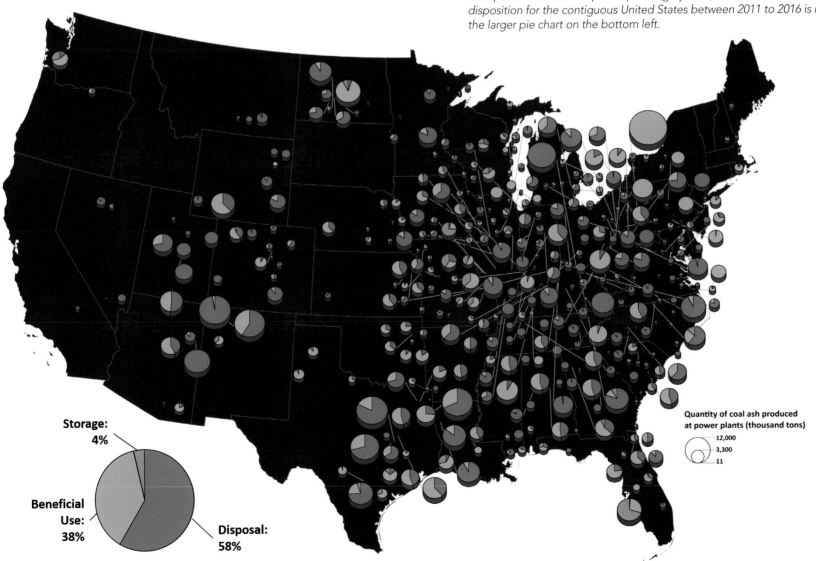

Storage: 4%

Beneficial Use: 38%

Disposal: 58%

Quantity of coal ash produced at power plants (thousand tons)

12,000
3,300
11

MOVING TOWARD A VIRTUAL SUBSURFACE

Coupling geoscience and GIS has helped scientists apply data-driven, systematic frameworks to better explore, characterize, analyze, and visualize the subsurface. This integration has provided new insights to enhance geo-discovery, improve the identification and prevention of geo-hazards, as well as offered solutions to ensure safe and enduring access to geo-resources. But more work must be done to fill in our knowledge gaps regarding the complex, dynamic subsurface system and our interactions with it.

Looking forward, new insights into the subsurface appear on the horizon because of the rapid evolution of technological advances in the fields of geoscience, GIS, and data science. These advances include progress in data manipulation, integration, analysis, and visualization that are unlocking powerful GIS, machine learning, and AI solutions to foster new opportunities to "see" into the subsurface. Further pairing

of the geosciences with GIS, machine learning, and advanced computing improve methods to overcome numerous time-consuming challenges encountered because of the large, unstructured, dispersed, and uncertain nature of subsurface data. These integrated solutions will enable geoscientists to move up the data science pyramid faster, thereby arriving at new discoveries, insights, and solutions that improve our predictions of subsurface properties and optimize how we interact with the subsurface.

Together, geoscience, GIS, machine learning, and advanced computing can help us build a more comprehensive, virtual understanding of our subsurface by combining disparate data in new and powerful ways. Ultimately, they allow us to further and more clearly peer beneath the basemap to understand the dynamic, complex subsurface system of our world.

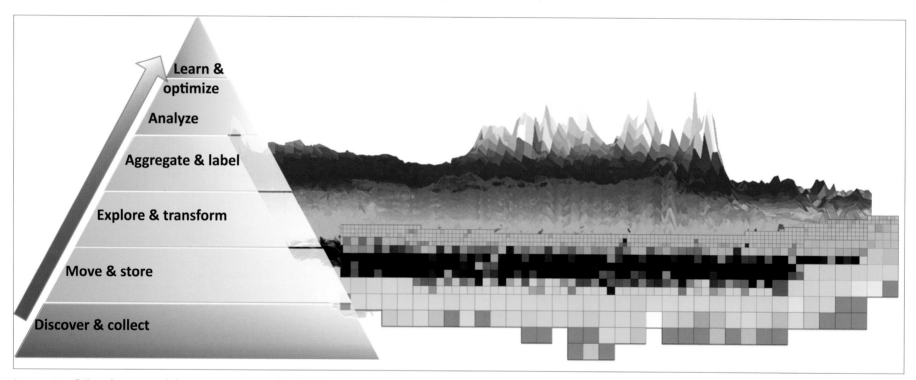

Integrating GIS and geospatial data, geoscience tools will help researchers move up the data pyramid and derive insights faster to improve human exploration of our planet's subsurface systems.

Acknowledgments

Parts of this technical effort were performed in support of the National Energy Technology Laboratory's ongoing research under the Carbon Storage Field Work Proposal DE FE-1022403, Coal Beneficiation Field Work Proposal DE FE-1022432, Offshore Unconventional Resources Field Work Proposal DE FE-1022409, and Rare Earth Element Field Work Proposal DE-FE-1022420 by NETL's Research and Innovation Center, including work performed by Leidos Research Support Team staff under the RSS contract 89243318CFE000003.

Disclaimer

NOTES

1. Owen, Edgar Wesley, "Oil Finders: A History of Exploration for Petroleum" (Washington, D.C.: American Association of Petroleum Geologists, 1975), 12.

2. Caplinger, M. W. "Allegheny National Forest Oil Heritage, Historic American Engineering Record," HAERNo.PA-436 National Park Service, US Department of the Interior (1997).

3. Rose, Kelly Kathleen, "Signatures in the Subsurface-Big & Small Data Approaches for the Spatio-Temporal Analysis of Geologic Properties & Uncertainty Reduction," PhD diss., Oregon State University (2016).

4. Cochener, John, "Quantifying drilling efficiency," US Energy Information Administration Office of Integrated Analysis and Forecasting (2010), https://www.eia.gov/workingpapers/drilling_efficiency.pdf.

5. Energy Data eXchange (EDX): https://edx.netl.doe.gov.

6. Geocube: https://edx.netl.doe.gov/geocube.

7. OpenEI: https://openei.org/datasets/dataset.

8. Homeland Infrastructure Foundation-Level Data (HIFLD): https://hifld-geoplatform.opendata.arcgis.com/.

9. GeoPlatform: https://www.geoplatform.gov.

10. Rose, Kelly, J. Bauer, V. Baker, A. Bean, J. DiGiulio, K. Jones, D. Justman, R. M. Miller, L. Romeo, M. Sabbatino, and A. Tong, "Development of an Open Global Oil and Gas Infrastructure Inventory and Geodatabase," NETL-TRS-6-2018, NETL Technical Report Series, US Department of Energy, National Energy Technology Laboratory (2018), DOI: 10.18141/1427573.

11. Morkner, Paige, C. G. Creason, M. Sabbatino, P. Wingo, J. DiGiulio, K. Jones, R. Greenburg, J. Bauer, and K. Rose, "Distilling Data to Drive Carbon Storage Insights," *Computers & Geoscience* (under review).

12. MechanicalSoup: https://github.com/MechanicalSoup/MechanicalSoup.

13. PySpider: http://docs.pyspider.org/en/latest.

14. Scrapy: https://scrapy.org.

15. Selenium: https://pypi.org/project/selenium.

16. Romeo, Lucy, P. Wingo, J. Nelson, J. Bauer, and K. Rose. 2019. "Cumulative Spatial Impact Layers," National Energy Technology Laboratory's Energy Data eXchange Tools (2019), https://edx.netl.doe.gov/dataset/cumulative-spatial-impact-layers, DOI: 10.18141/1491843.

17. Rose, Kelly, J. R. Bauer, and M. Mark-Moser, "A Systematic, Science-Driven Approach for Predicting Subsurface Properties," *Interpretation*, 8, no. 1 (2020): 167-181, https://doi.org/10.1190/INT-2019-0019.1.

18. Mark-Moser, M., R. Miller, K. Rose, J. Bauer, and C. Disenhof, *Detailed Analysis of Geospatial Trends of Hydrocarbon Accumulations, Offshore Gulf of Mexico,* NETL-TRS-13-2018, NETL Technical Report Series, US Department of Energy, National Energy Technology Laboratory (2018), DOI: 10.18141/1461471.

19. Rose, Kelly, C. G. Creason, D. Justman, and R. Thomas, "Developing a Geo-Data Science Driven Method to Assess REE's in Coal Related Strata" (2018 Annual Project Review Meeting for Rare Earth Elements Program. Pittsburgh, PA, April 2018), https://netl.doe.gov/sites/default/files/netl-file/20180410_1100C_Presentation_Rose_NETL.pdf.

20. Davies, Richard, G. Foulger, A. Bindley, and P. Styles, "Induced Seismicity and Hydraulic Fracturing for the Recovery of Hydrocarbons," *Marine and Petroleum Geology,* 45 (2013): 171-185.

21. Justman, Devin, K. Rose, J. Bauer, R. Miller, and V. Vasylkivska, "Evaluating the Mysteries of Seismicity in Oklahoma" (2015), https://edx.netl.doe.gov/dataset/edit/oklahoma-seismicity-story-map.

22. Sabbatino, Michael, L. Romeo, V. Baker, J. Bauer, A. Barkhurst, A. Bean, J. DiGiulio, K. Jones, T. J. Jones, D. Justman, R. Miller III, K. Rose, and A. Tong, "A Global Oil & Gas Features Database" (2017), https://edx.netl.doe.gov/dataset/global-oil-gas-features-database, DOI: 10.18141/1427300.

23. Styron, Richard, "Gemsciencetools/gem-global-active-faults: First Release of 2019," *Zenodo* (2019), doi:10.5281/zenodo.3376300.

24. Oklahoma Geological Survey, "Oklahoma Earthquake Catalog, 1970–2016" (2016), http://www.ou.edu/ogs/research/earthquakes/catalogs.

25. Holland, Austin A., "Preliminary Fault Map of Oklahoma," OF3-2015, *Oklahoma Geological Survey Open-File Report* (2015).

26. Peacock, D. C. P., V. Dimmen, A. Rotevatn, and D. J. Sanderson, "A Broader Classification of Damage Zones," *Journal of Structural Geology,* 102 (2017): 179-192.

27. Justman, Devin, C. G. Creason, K. Rose, and J. Bauer, "A Knowledge-Data Framework and Geospatial Fuzzy Logic-Based Approach to Model and Predict Structural Complexity," *Journal of Structural Geology* (under review).

28. Arbenz, J. K., "Structural Cross Sections OK3 and OK4 across the Oklahoma Ouachita Mountains," *OGS Circular* 112A (2008), http://ogs.ou.edu/docs/circulars/C112AP7.pdf.

29. Environmental Protection Agency, "Global Greenhouse Gas Emissions Data," Accessed March, 2020, https://www.epa.gov/ghgemissions/global-greenhouse-gas-emissions-data.

30. National Risk Assessment Partnership (NRAP): https://edx.netl.doe.gov/nrap/. Bauer, Jennifer, C. Rowan, A. Barkhurst, J. Digiulio, K. Jones, M. Sabbatino, K. Rose, and P. Wingo, "Natcarb" (2018), https://edx.netl.doe.gov/dataset/natcarb, DOI: 10.18141/1474110.

31. Laske, G. and G. Masters, "A Global Digital Map of Sediment Thickness," *EOS Trans. AGU*, 78, F483 (1997).

32. Straume, E.O., C. Gaina, S. Medvedev, K. Hochmuth, K. Gohl, J. M. Whittaker, J. M., et al. (2019). "GlobSed: Updated Total Sediment Thickness in the World's Oceans," *Geochemistry, Geophysics, Geosystems*, 20 (2019), DOI: 10.1029/2018GC008115.

33. East, Joseph A., *Coal Fields of the Conterminous United States-National Coal Resource Assessment Updated Version*, No. 2012-1205, US Geological Survey (2013).

34. Energy Information Administration, "Form EIA-860: Annual Electric Generator Report," US Energy Information Administration (2011-2016).

35. Energy Information Administration, "Form EIA-923: Power Plant Operations Report Instructions," US Energy Information Administration (2011-2016).

Geologic map on page 58 by Galloway, W. E. Chapter 15: "Depositional Evolution of the Gulf of Mexico Sedimentary Basin." In *Sedimentary Basins of the World*, Vol. 5: *The Sedimentary Basins of the United States and Canada*; Elsevier: The Netherlands, 2008.

PART 2
HOW EARTH LOOKS

How Earth looks is essentially how we as humans change Earth's appearance and function, as illuminated by linkages between natural science and social science, in science partnerships that work across disciplines, geographies, and organizations. Here, we often use GIS to interactively and iteratively create and evaluate alternative (geo)designs to make better decisions, especially with land cover for land-use planning, green infrastructure planning, urban planning, and sustainability science.

This GIS view of downtown Boston simulates shadow patterns at 1:40 p.m. on January 9, 2019. Try the interactive web app created by the Office of GIS, Boston Planning and Development Agency, at GISforScience.com.

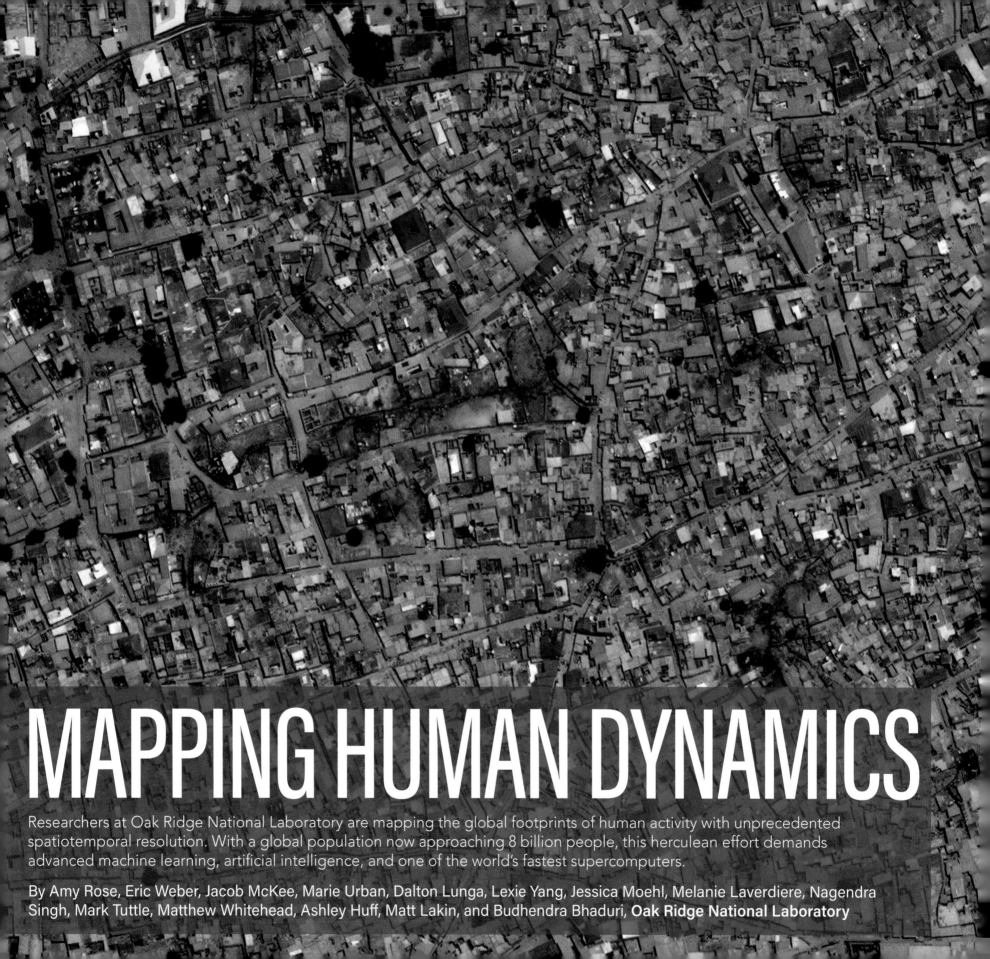

MAPPING HUMAN DYNAMICS

Researchers at Oak Ridge National Laboratory are mapping the global footprints of human activity with unprecedented spatiotemporal resolution. With a global population now approaching 8 billion people, this herculean effort demands advanced machine learning, artificial intelligence, and one of the world's fastest supercomputers.

By Amy Rose, Eric Weber, Jacob McKee, Marie Urban, Dalton Lunga, Lexie Yang, Jessica Moehl, Melanie Laverdiere, Nagendra Singh, Mark Tuttle, Matthew Whitehead, Ashley Huff, Matt Lakin, and Budhendra Bhaduri, **Oak Ridge National Laboratory**

With a population approaching 3 million people, Kano is Nigeria's second-largest city. It attracts migrants from throughout the northern part of the country. Its population grows every single day, and no reliable census exists. Not knowing how many people there are, or who they are, or where they live are challenges confronting humanitarian efforts in crowded cities like Kano.

INTRODUCTION

The critical need to account for vulnerable populations has shaped the pioneering population modeling program at Oak Ridge National Laboratory (ORNL) from the onset. The program's history in Tennessee is tied to ORNL's nuclear legacy as a Manhattan Project site, initially established to support US atomic weapons priorities in the 1940s. The program matured into a US Department of Energy national laboratory with leading expertise in peaceful applications of nuclear energy, including power production, medical isotopes, and neutron-scattering research for open science. Population and environmental risk assessment developed alongside the lab's evolving missions, beginning with the need to examine local contamination and exposure risks and becoming a multifaceted program that today looks at populations globally.

Basic questions of how many people are at risk and where they are located continue to motivate novel approaches to mapping dynamic human populations that grow, decline, and move at often unpredictable paces. ORNL's current population research combines geospatial and computational expertise to model spatially and temporally explicit populations at very high resolution. ORNL uses scalable methods to assess any region of the world and locate groups most at risk for or affected by the varying insecurities of urban settlement, ranging from natural disasters, infectious diseases, and strained resources to rapid population growth, migration, and sudden displacements.

Data and compute advances during the past two decades have brought greater accuracy and immediacy to the work of locating global populations. The enormous volume of existing high-resolution satellite imagery along with rapidly increasing refresh rates enable the detection of formal and informal settlements everywhere in the world. The parallel-computing power of today's graphics processing unit (GPU) systems has dramatically accelerated the time needed to extract relevant information from large datasets. Settlement maps can now capture abrupt changes in built-up areas, such as the overnight appearance of makeshift refugee camps, a feat that was not easily achievable before the 2000s.

The ORNL challenge—distributing nearly 8 billion people into the trillions of pixels estimated to contain the world's current population—is no small proposition.

Since the 1990s, the ORNL group has undertaken a critical effort to map and estimate the world's population at a global scale. Their modeling-based approach combines high-resolution imagery, statistical data, and computational resources to support US and global population distribution databases with 90-meter spatial resolution and variable temporal resolutions.

The program's earliest questions—where, when, and how many—are now evolving to capture the dynamics of mobile populations, enabling new insights to inform urban development, socioeconomics, humanitarian campaigns, and emergency management.

ORNL uses a full suite of geographic information system (GIS) mapping expertise and resources to deepen understanding of populations in motion, moving beyond where people are to the nuances of who they are, why they move, where they go, and what they do. Their efforts expand, on a global scale, possibilities for how GIS data and tools can be used to locate volatile at-risk groups before, during, and after a crisis.

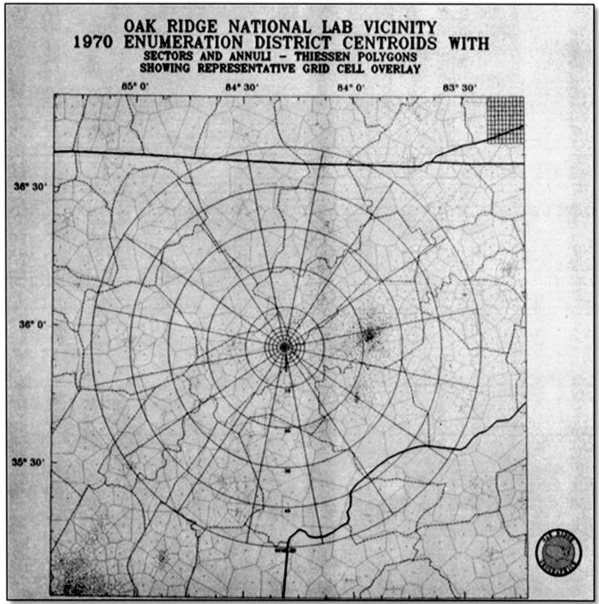

ORNL population risk-assessment maps produced in the 1970s with US Census input data.

NORTHEAST U. S. (PJM REGION)
POPULATION DENSITY
FINAL 1980 CENSUS

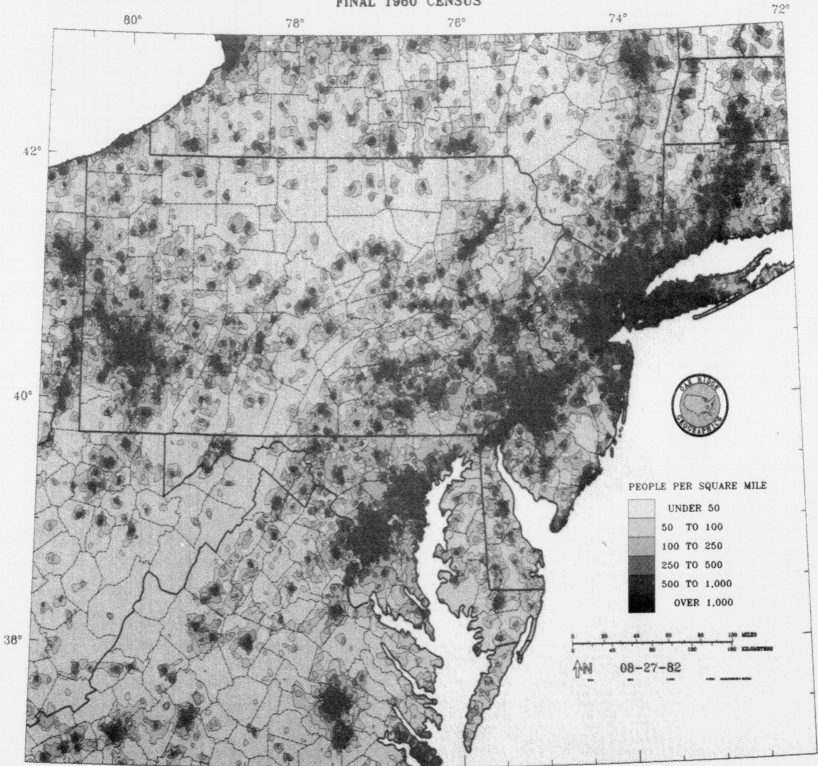

PEOPLE PER SQUARE MILE

UNDER 50
50 TO 100
100 TO 250
250 TO 500
500 TO 1,000
OVER 1,000

08-27-82

This pen-plotted population density map—created for the final 1980 Census using an early GIS system developed in-house with Fortran using Census TIGER geography—shows the Northeast United States when the total US population was 226.5 million (compared to 330 million in 2020).

NIGERIA: A BOTTOM-UP APPROACH

In 2013, ORNL began a critical effort to locate previously unmapped populations in northern Nigeria as part of a multiyear collaboration with the Global Polio Eradication Initiative (GPEI). The project demonstrated the need for a census-independent, bottom-up approach to mapping and estimating populations with the spatial precision needed to account for specific demographic subpopulations, especially in data-poor countries.

The GPEI has conducted vaccination campaigns aimed at children younger than age five since the 1980s and successfully eradicated polio viruses in much of the developing world, but the disease remains endemic in three countries: Afghanistan, Nigeria, and Pakistan. Limited access to at-risk communities and inadequate information on target populations can undercut the effectiveness of vaccination efforts in the most vulnerable areas.

Nigeria's most recent national census in 2006 was well out of date at the time of the project, and the information it provided at the local government level made it difficult to identify smaller communities within these administrative areas. Moreover, projected population estimates were based on constant growth rates that missed the accelerated development of Nigeria's urban areas, leaving more than a million people essentially invisible from any records.

To locate vaccine-eligible children in Nigeria, ORNL developed a model-based approach incorporating layers for settlement areas, building types, and population density. Researchers used supervised machine learning to extract settlement areas from high-resolution satellite imagery and classify the results into residential and nonresidential categories. Microcensus surveys conducted by locals provided per-building population counts in sample locations. Combined with the classified settlement layer, population density estimates from the survey data informed residential population counts at a fine spatial (<100 meters) resolution.

The geospatial dataset created for the Nigeria effort helped identify chronically missed settlements in polio vaccination campaigns and continues to play a critical role in eliminating polio from the developing world. The initial GPEI-based project expanded to support other world health missions in sub-Saharan Africa and South Asia.

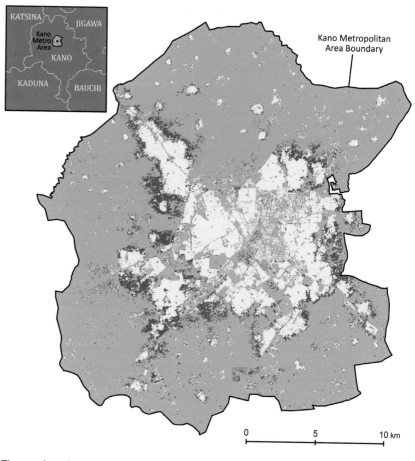

The total settlement area in Kano, Nigeria, increased more than 40% from 2006 to 2014. Estimates based on past trends projected a 2.02% annual increase in built-up areas, while ORNL's mapping results showed an increase of 4.37% per year. This image juxtaposes 2006 settlement areas (tan) with additional areas detected in 2013 imagery (red). Eric M. Weber et. al, "Census-Independent Population Mapping in Northern Nigeria," Remote Sensing of Environment.

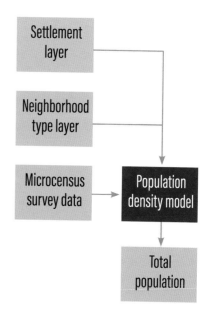

ORNL's approach captured accelerated growth post-2006 in Kano, Nigeria, accounting for new, unmapped settlements and yielding a population estimate much higher than previously published projections.

SOKOTO

KATSINA

YOBE

ZAMFARA

JIGAWA

BORNO

KANO

BAUCHI

GOMBE

KADUNA

NIGER

ADAMAWA

N I G E R I A

PLATEAU

ABUJA

KWARA

NASSARAWA

OYO

TARABA

EKITI

KOGI

OSUN

BENUE

OGUN

ONDO

EDO

LAGOS

EBONYI

ANAMBRA

ENUGU

DELTA

IMO

ABIA

AKWA IBOM

RIVERS

BAYELSA

*Kano City is located in Kano State,
a jurisdiction in northern Nigeria.*

METHODS

ORNL's modeling-based approach to mapping human dynamics incorporates settlement, contextual, and population layers into geospatial datasets that can provide building-level population insights for anywhere in the world. Observations at the level of individual buildings now offer context on land use, neighborhood type, occupancy, and demographics. Population estimates are time variant and mapped at pixel level across the globe.

- Level 1—Settlement layer: building feature-extraction methods are applied to high-resolution imagery to detect individual buildings and map land use in built-up areas.
- Level 2—Contextual layer: urban land use is characterized using sampling workflows and automated feature learning techniques.
- Level 3—Population layer: novel statistical methods are used to estimate population density and capture human activity patterns globally.

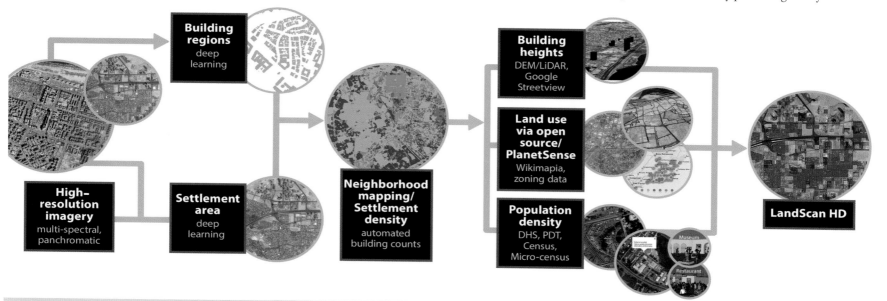

Imagery processing pipeline

The vast amounts of data required for ORNL's mapping projects travel a long road of processing before researchers can begin their study. As an example, a query for images of Washington State yields 11,435 image strips—swaths of high-resolution satellite images, each covering an average land area of 1,200 km². A second search filters out images obscured by clouds or otherwise judged unsuitable, whittling the results to 605 strips of ground images. Those results add up to a total of 2.7 terabytes—2.7 trillion bytes, or the equivalent of around three-quarters of a billion pages of text. High-speed data transfer nodes at ORNL then move the data to servers and begin parallel processing the images. This stage of the operation decompresses the data, corrects for perspective and terrain, and sharpens the resolution, ballooning the size to 26 terabytes, or nearly 10 times the original. A single processed image may consist of more than 2 billion pixels. To support GIS population research, the Compute and Data Environment for Science (CADES) system at ORNL holds about 1.5 million images at any given time, a total of about 2.6 petabytes of data.

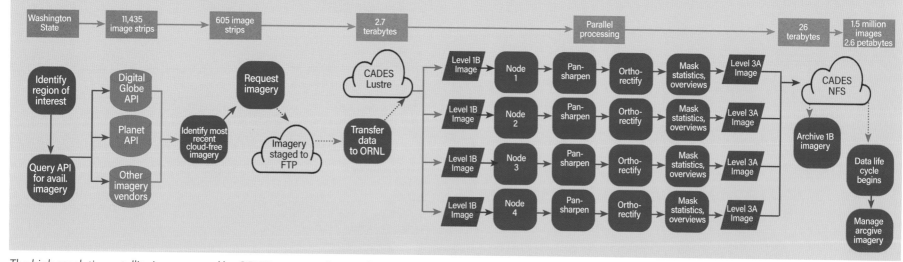

The high-resolution satellite imagery used by ORNL's geospatial researchers translates to massive amounts of data. Researchers once physically transported the data via piles of hard drives but now conduct preprocessing transfers through the cloud.

From settlement mapping to building feature extraction

Historically, ORNL developed settlement mapping methods to analyze high-resolution imagery and distinguish settled and unsettled areas. The output achieved a general representation of built-up areas, initially at a coarse, 8-meter resolution. Today the program uses building feature extraction methods to describe individual buildings at near half-meter resolution. The evolution from binary to multilayered, contextual information on the structure and type of individual buildings has expanded opportunities for data analysis and enabled novel outputs relevant to the timely and specific needs of user communities.

Advances from many directions, especially imagery, sensing, and computing, continue to stimulate progress. In recent years, GPU-based platforms have become smaller and more affordable, making HPC (high-performance computing) capabilities available to data analysts and not limited to supercomputing facilities. Although ORNL's leadership-class computing systems, including the 200-petaflop supercomputer Summit, inform the development of population modeling, day-to-day processing happens on mini-GPU clusters that bring parallel-computing power to desktops.

Foundational imagery needed for settlement detection has evolved in resolution and availability. The satellite-based data that originally fueled settlement mapping were mostly panchromatic images physically transported to ORNL on hard drives. Now multispectral input imagery can be transfered on demand via cloud servers. Image quality and global coverage make satellite data essential to ORNL's large-scale applications. However, advances in remote sensing and drone technologies are dramatically improving image resolutions—from meters to centimeters—and are potential sources of new information.

Machine-learning algorithms for object detection, image segmentation, and classification have also been enriched by technological advances. The algorithms developed at ORNL in the 1990s during the early days of the population-modeling program remain fundamental to current approaches. The difference now is that models have more and better imagery for training data, and outputs come in hours rather than months, opening realistic routes to improve algorithms.

Upgraded compute power and advanced algorithms now make it possible to rapidly extract building features from large datasets, such as countries and groups of countries with input imagery reaching hundreds of terabytes. For example, processing the country of Nigeria in 2014 took between four and five months. Processing Afghanistan in 2019 took six days. The speed boost stems from parallel processing on GPUs and evolved modeling approaches generalizable to large areas. Early modeling relied on support-vector machine learning that required building multiple unique models to cover an area as large as a country. In addition to longer processing times, more models require more human effort to manually label training data. Now, machine-learning algorithms can accommodate variations in landscape types, such as distinguishing forested terrain from deserts, increasing the workload one model can handle. A single model can be used to extract building features from an entire country.

A standout feature of the ORNL program is its scale of commitment to a mission that drives GIS technologies toward future, sustainable routes for modeling populations globally. Creating efficient models to extract building features from massive imagery archives—truly big data—over very large land areas is one step. However, the program encompasses all the steps that follow, covering current bottlenecks as well as emerging challenges.

The vision for optimizing the process of settlement and building detection for global-scale applications is essentially a quest for practical methods to store and analyze the estimated 10 trillion pixels that cover the planet's land area—and maintain the output.

Researchers describe the challenge as akin to "drinking from a fire hose." More imagery comes in than they can realistically use. Significant labor goes into the immediate work of making the imagery viable for analysis so that its value becomes tangible. Looking ahead, sights are set on greater automation and efficient methodologies to keep the output current.

As an example, in 2021, ORNL will complete a baseline dataset for all US structures larger than 450 square feet. This dataset includes regularized structure outlines with a variety of attributes attached to each structure, including occupancy type, address, and height. The project began in 2017, meaning some of the input imagery comes from that date or earlier. Potential updates to the end product are less of a compute challenge than a pragmatic one. The computational horsepower and imagery needed to rebuild models from the ground up already exist, but is the exercise practical or efficient? Future strategies that use machine learning to filter essential information from large image files, determine where changes occur, and automate updates are all part of a sustainable, scalable approach to population modeling.

Advances in the spatial resolution of source data improve observations of the built environment. The comparison here shows urban areas detected for the same area using satellite imagery of varying spatial resolutions, including MODIS (500 meters) and Landsat (30 meters). ORNL's Settlement Mapping Tool (SMT) achieves high spatial precision and granularity at half-meter resolution using WorldView-1 satellite imagery (0.46 meters).

HIGH-PERFORMANCE COMPUTING FOR POPULATION MODELING

ORNL is home to the nation's most advanced supercomputing resources for open science. The lab's leadership-class computing resources have included Jaguar, which became the science community's first petaflop system in 2005, followed by the 20-petaflop Titan supercomputer in 2012. At 10 times more powerful than its predecessor, Titan introduced a hybrid GPU-CPU architecture and operated as a top-ranked system supporting researchers from all over the world until it was retired in 2019. ORNL's current flagship computer Summit boasts 200 petaflops at peak performance.

ORNL's HPC capabilities enable researchers to develop machine-learning workflows and prototype models, as well as explore potential new directions on HPC architectures.

In 2017, ORNL researchers were allocated 25,000,000 processor hours on Titan for a project exploring HPC-accelerated approaches to settlement mapping. The team processed more than 45 terabytes of imagery for Yemen in less than two hours. New projects are planned on Summit in 2020, a machine with eight times the computational power of Titan.

First fired up in 2005, the Jaguar supercomputer was built at Oak Ridge Lab by Cray and had 224,256 x86–based AMD Opteron processor cores (and operated with a version of Linux called the Cray Linux Environment).

Summit is an IBM supercomputer designed for use at ORNL. As of November 2019, it is one of the fastest supercomputers in the world, capable of 200 petaflops. It is also one of the world's third most energy-efficient supercomputers, with a measured power efficiency of 14.668 gigaflops per watt.

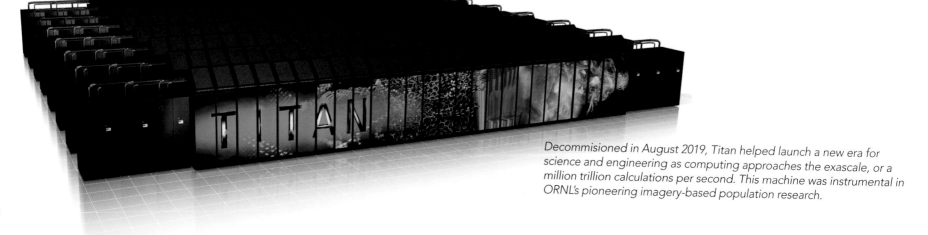

Decommisioned in August 2019, Titan helped launch a new era for science and engineering as computing approaches the exascale, or a million trillion calculations per second. This machine was instrumental in ORNL's pioneering imagery-based population research.

NEIGHBORHOOD MAPPING — BUILDINGS WITH CONTEXT

Building-feature extraction methods generate first-order (L1) information about settlement areas, such as individual building footprints. Once buildings are detected, researchers can map urban land use (L2) and characterize neighborhoods or clusters of buildings according to use functions, ranging from industrial to residential categories. Characterization provides additional layers of knowledge that open pathways for targeted analysis. For example, humanitarian campaigns may need to locate populations in impoverished areas, so maps that highlight informal settlements are critical to success. By distinguishing densely populated areas, impoverished areas, or industrial parks, neighborhood mapping generates contextual information about a city's settlements with far-reaching potential. Understanding the spatial arrangement of neighborhoods can point to vulnerabilities and support many initiatives to fight infectious diseases, stimulate economies, expand access to resources, and otherwise sustain urban communities.

To map land-use patterns at scale, machine-learning algorithms are applied to high-resolution imagery to generate contextual information over very large areas. As a comparative example, large visual databases of global imagery, e.g., ImageNet, cover in their entire library only a fraction of the land surface area ORNL typically analyzes for a single country. One of the biggest challenges to scaling the approach is fitting the models to enormous datasets with extensive diversity. Computers learning to identify industrial parks in South America suddenly need to detect industrial areas in Southeast Asia that look very different. The question becomes how best to develop algorithms that can accommodate differences in terrain, land-use patterns, architectural style, structural density, and other variations across the globe. Packing in more and more training imagery with additional cultural diversity eventually results in underfit models that perform poorly. Thus, for large-scale mapping projects, training models to recognize building categories inevitably tests the limits of how much input data the approach requires and can handle.

ORNL uses experience-based machine learning to map neighborhoods. The process involves gathering imagery that contains all the neighborhood types the computer needs to detect in a given geography. Next, deep-learning algorithms are designed to examine the training data and encode the patterns that distinguish building categories. Having the computer figure out the distinguishing features of a neighborhood, e.g., industrial and residential areas, avoids the hand-engineering otherwise required to manually code all the rules that define each category. Deep-learning advances such as this add efficiency that makes large-scale mapping feasible. The approach likewise gains additional accuracy in instances where computers detect subtle features that may be difficult for humans to interpret.

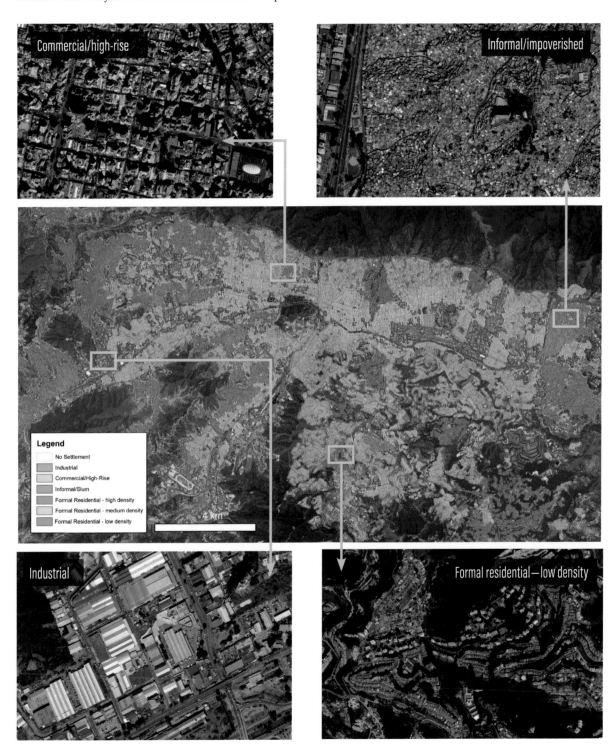

Neighborhood mapping workflows applied to satellite imagery in Caracas, Venezuela, identify building categories to generate insights on the socioeconomic patterns of urban land use.

Getting a good fit

Given the scope of the ORNL program, which aims to map every pixel across the globe, the organization will need more than one model to meet the challenge. Ideally, the fewer the models there are, the better. Current research is testing a proof-of-concept workflow with multiple models that characterize sample imagery across the planet. The concept is based on observations that pockets of similarities tend to reappear globally that are not necessarily tied to regional geographies. The idea is to analyze the entire collection of global imagery using an algorithm that can gather similar pixels and distribute them into multiple "buckets" for unique models. When unseen imagery enters the workflow, the computer quickly determines which bucket to use and pushes the input through the appropriate model. The beauty of the highly automated workflow is that once the buckets are allocated, the corresponding models can be changed as needed to support not only neighborhood mapping but also other kinds of analysis.

Estimating global populations—a people-per-pixel approach

To populate settlement and contextual layers at scale with the critical "missing pixels"—people—researchers must overcome global disparities in data availability. While the United States conducts regular censuses and household surveys that provide population insights at the national level, many other countries do not record adequate, up-to-date, or reliable population information (if any at all) at the minimum coverage needed to estimate and distribute populations with any degree of confidence. That means outside of North America, top-down disaggregation methods of population modeling are impractical. To scale population modeling for the world, bottom-up solutions for aggregating populations in any country are essential.

A unique feature of the ORNL program is the use of statistical data in addition to or in lieu of census reporting to model populations outside the United States. ORNL-developed population density tables (PDT) report building occupancy estimates of people per 1,000 square feet at the national and regional levels and for night, day, and episodic activities where large gatherings occur. Using a Bayesian statistical machine-learning approach, baseline models cover every geographic area in the world and include more than 50 building functions, ranging from residential households to museums, churches, schools, hospitals, and even open-air locations people visit such as cemeteries. The PDT database is dynamically updated and published through a content-management system that reports building occupancies in both tabular and geospatial formats.

The concept is about capturing human activity at the building level to understand how people use spaces in normal patterns of life—during a typical workweek, weekends, holidays, special events, and potential seasonal fluctuations.

Mining the source data is an exhaustive attempt to collect snapshots of all these experiences. The project includes personal accounts, subject-matter experts, and an assemblage of open-source data such as publications, websites, and reports.

Open-source data

PDT collects open-source data from more than 50,000 published references, including:

- Academic journals
- Official government statistics
- Corporate and university webpages
- Tourism brochures
- Nongovernmental organization publications and data
- Real estate databases
- Surveys
- Websites
- Images

The database delivers transparent information about the source data, methodology, and uncertainty for probabilistic population density estimates. All source data are geotagged and reviewable.

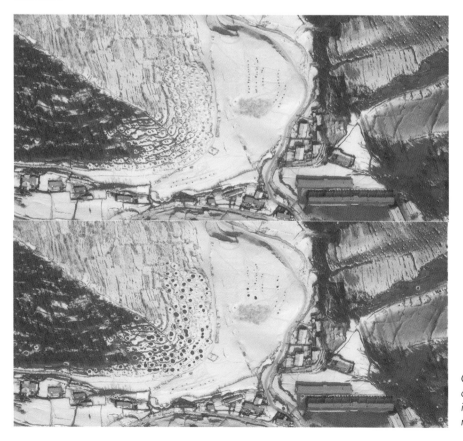

ORNL is testing the feasibility of automating the identification and quantification of graves—both formal and informal—like these in North Korea. By using satellite imagery and modified convolution neural networks for image classification, researchers can automate the manual quantification of graves.

TIME VARIANCE

ORNL's population-modeling approach achieves high temporal resolution in addition to fine spatial resolution. Ambient, day, and nighttime estimates are mappable for populations in and outside of the United States. A broad distinction between day and night values for residential populations can be thought of as ranges when daytime populations are likely at work, school, or moving through daily routines, versus nighttime ranges, when populations are expected to be at their residences. Not all countries have the same day and night ranges because business hours and cultural activity patterns vary around the world. US Census data inform time-variant population estimates for America, while PDT informs estimates globally. As more data become available, the day and night distinction may evolve into a 24-hour account of the world's activity patterns.

For the United States, ORNL has created LandScan USA, the only national dataset delivering day and night residential populations at an incredibly high 90-meter resolution.

The capability to deliver time-variant population information has been critical to national risk assessment and emergency management. As a unique geospatial resource, LandScan USA has proved indispensable to the Federal Emergency Management Agency (FEMA), Department of Homeland Security, and National Geospatial-Intelligence Agency missions.

San Francisco daytime population

New York City daytime population

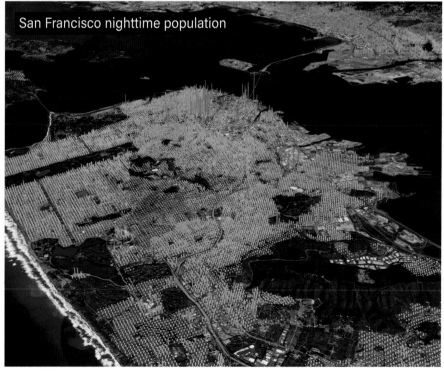

San Francisco nighttime population

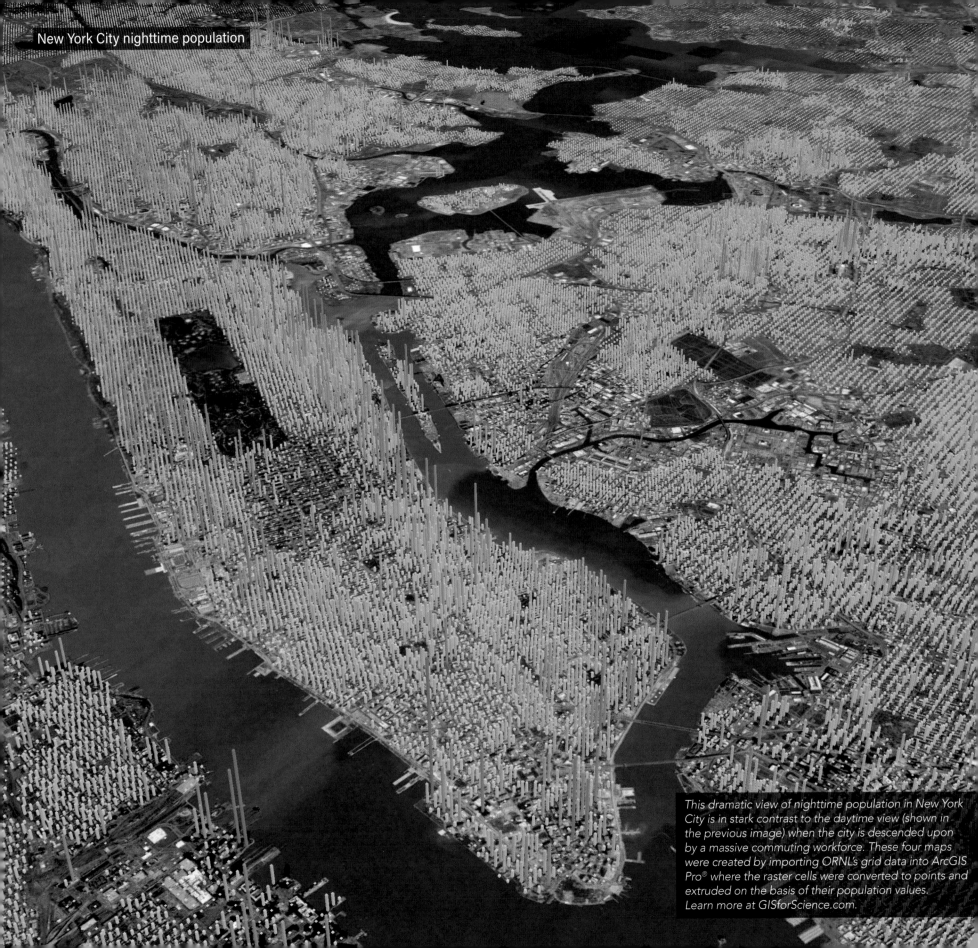

New York City nighttime population

This dramatic view of nighttime population in New York City is in stark contrast to the daytime view (shown in the previous image) when the city is descended upon by a massive commuting workforce. These four maps were created by importing ORNL's grid data into ArcGIS Pro® where the raster cells were converted to points and extruded on the basis of their population values. Learn more at GISforScience.com.

IMPACTS

Critical infrastructure and emergency response

ORNL develops critical infrastructure data that supports US federal agencies in assessing the interdependent vulnerabilities of at-risk populations and infrastructures during emergencies. Open-source data are used to create critical infrastructure layers that inform ORNL's population models and contribute to user platforms, namely, the Homeland Infrastructure Foundation-Level Data open platform, a public domain resource for geospatial data to support preparedness, resiliency, and research among diverse user communities. National critical infrastructure includes schools, prisons, rail networks, day care facilities, solid waste facilities, mobile home parks, hospitals, energy infrastructure (e.g., petroleum, natural gas, electricity), major sports venues, national shelters, nursing homes, law enforcement, and convention centers.

ORNL data supported emergency response during the record 2017 hurricane season that impacted areas of Texas, Florida, Puerto Rico, the Virgin Islands, and Caribbean territories.

As an example, to address Hurricane Harvey–related flooding in Texas, ORNL delivered buildings and structures data for the state's coastal counties—processing 2,000 images covering 26,000 square miles of land—in just 24 hours. Natural disasters and other emergencies have intensified the need to quickly identify and characterize vulnerable and affected populations. To that end, ORNL makes invaluable data contributions to FEMA and other government agencies, enabling first-response efforts as well as postimpact damage assessments. A massive effort in 2017 supported federal response to areas of Texas, Florida, Puerto Rico, the Virgin Islands, and Caribbean territories in the wake of Hurricanes Harvey, Irma, and Maria. In recent years, ORNL has provided critical, timely population and infrastructure data to help assess the impacts of devastating volcanic eruptions in Hawaii and raging wildfires across California.

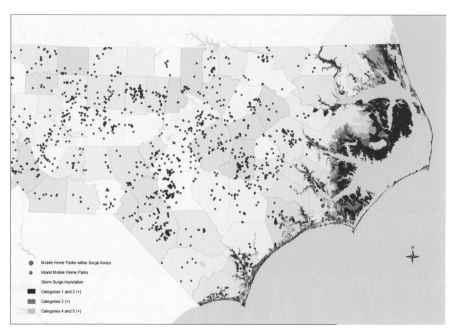

Mobile homes, which are especially vulnerable to tornadoes and other natural disasters, were previous gaps that are now included in national infrastructure data.

Acknowledgment

This manuscript has been authored by UT-Battelle, LLC under Contract No. DE-AC05-00OR22725 with the US Department of Energy. The United States Government retains and the publisher, by accepting the article for publication, acknowledges that the United States Government retains a nonexclusive, paid up, irrevocable, worldwide license to publish or reproduce the published form of this manuscript, or allow others to do so, for United States Government purposes. The Department of Energy will provide public access to these results of federally sponsored research in accordance with the DOE Public Access Plan. The authors would like to acknowledge the funding support from US Government Agencies and the Bill and Melinda Gates Foundation for the research discussed here.

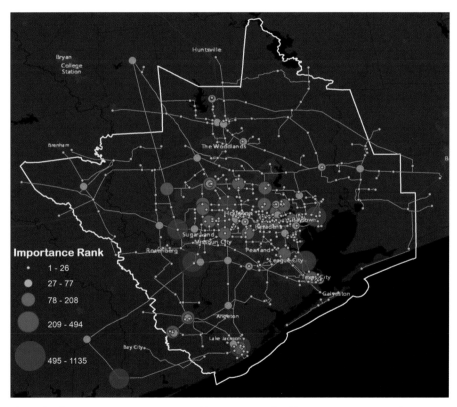

Critical infrastructure across Houston, Texas, was ranked by concentration and importance and combined with population data to identify populations at the highest risk during Hurricane Harvey–related flooding.

SUSTAINABLE FOOD PRODUCTION

Facing the prospect of feeding an additional 2 billion people by the year 2050 has agricultural scientists scrambling for practical and sustainable solutions. Using data from a broad range of sources, geospatial innovations are creating breakthroughs.

By Paul West, James Gerber, and Deepak Ray, **University of Minnesota, Institute on the Environment;** and Mauricio Castro Schmitz, **The Nature Conservancy**

Small plot vegetable farming in Ethiopia. Small and medium farms occupy more global land than large farms and provide much of the world's crop diversity. The harvests from small farms also tend to be consumed closer to their source.

CREATING A SUSTAINABLE GLOBAL FOOD SYSTEM

One of humanity's grand challenges is feeding a growing population on a warming planet. The current situation is daunting—hundreds of millions of people go hungry most days. And while crop production must double between 2010 and 2050 to bridge the gap,[1] current yield projections are not on track to meet the anticipated demand.[2] Agriculture already occupies about 38% of the ice-free land,[3] including the best land for growing food. Further, agriculture accounts for 70% of the freshwater used by people[4] and 20% to 25% of all greenhouse gas emissions.[5]

Agriculture activities contribute to degraded water quality and are the leading driver of habitat loss globally, especially across the tropics. A changing climate, growing population, and increasingly rich diets accelerate these challenges.

With nearly three-quarters of the planet covered in water, a relatively small part of the earth is available for humans and other land-dwelling organisms; this map[3,6] depicts the current balance of Earth's croplands and pasture lands.

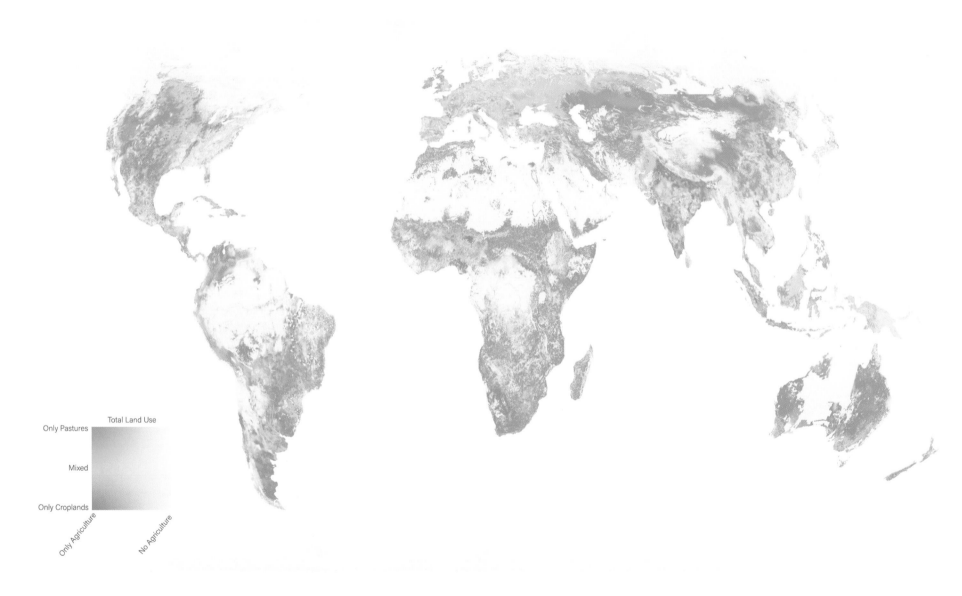

People use about 38% of Earth's ice-free land to grow crops and raise animals.[3,6] More than 15 million square kilometers, an area about the size of South America, are needed just to grow crops. For pasture and rangeland, we use more than 32 million square kilometers—an area comparable to the continent of Africa. In comparison, urban areas cover less than a half percent of total land area.

FOOD IS MORE THAN CALORIES

How can humanity rise to meet the global food supply challenges? This chapter explores how scientists at the University of Minnesota's Global Landscape Initiative (part of the Institute on the Environment) study the many complex geographic factors that interact to shape the global food system. Integrating and synthesizing geographically referenced data from a myriad of international, national, regional, and local sources form an inherently large-scale geospatial problem. Ecological factors, including climate, soils, topography, and geology, provide the basic constraints that determine what's possible for food production. Socioeconomic factors—prices, policies, cultural preferences, land tenure, management, etc.—add further complexity. Combined, these factors determine which food grows where, how it is produced, whether a crop becomes feed or fuel, how it impacts the environment, how the climate changes, who trades with whom, and where food waste occurs. By 2050, the world population is expected to reach about 10 billion. That's more than 2 billion more people who need food, water, and shelter to survive. This rapid population growth, combined with rising dietary and biofuel consumption, has led to a major transformation of Earth's land, water, and air systems.

Spatial data and computational analysis with a geographic information system (GIS) are essential for both assessing the challenge and designing solutions. The patterns, trends, opportunities, risks, and trade-offs in the food system vary across scales and from place to place. For example, the relationships among scales and stakeholders might look like this: multilateral institutions working at the global scale assess progress toward Sustainable Development Goals adopted by the United Nations. Other stakeholders, such as development banks, develop strategies designed to address regional strategies and identify projects in which investments can promote sustainable development and improve human well-being. Nationally, governments work with stakeholders to set goals and policies to meet their specific needs as well as achieve their targets toward meeting the Sustainable Development Goals, the Paris Accord, and other international agreements. Locally, communities design strategies in which farming is important for tradition and the economy, and environmental resources are a protected tradition and important to the economy. Having a common information platform on which to synthesize all these data is imperative. Ideally, the assessments and solution designs are integrated such that global assessments shape regional solutions, regional assessments shape local solutions, and local solutions help achieve local and global goals. While this ideal situation may not be common in practice, achieving it is possible only if spatial data are managed and used to make science- and place-based decisions.

Fortunately, many companies, governments, development banks, foundations, non-governmental organizations (NGOs), and others are working toward a sustainable food system. Their work includes reducing the environmental impacts of commodity production, increasing yields in regions where poverty is high and food is scarce, and adapting to a changing climate. Since the many ecological and social parts of the food system are all intertwined, strategies to improve food security must be holistic. In the absence of this approach, progress in areas such as increasing commodity production through irrigation may do little for (or further set back) efforts to improve local food and water availability.

This chapter examines several data-driven strategies for improving global food security and the environment through a three-part strategy: producing more food on current agricultural land, growing food sustainably, and using what we already grow more efficiently.

At the end of this chapter, a case study in Latin America illustrates how large-scale analyses and datasets can be integrated to direct action at regional and local scales.

Calorie Production on Croplands

Low   Moderate   High

Food not only provides dietary energy, it is also a source of many different nutrients that play important roles in human growth and development, as well as disease prevention and longevity. Getting all the nutrients we need to grow, develop, and thrive requires eating and producing a variety of foods.

THE BIG PICTURE: FARM SIZE AND NUTRIENT DIVERSITY

Across the planet, farms vary greatly in size and in what they grow. In many parts of the developing world, small farms play a critical role in producing a diversity of foods that are essential for local food security and nutrition. Collectively, small farms produce between half and three-quarters of the world's food and micronutrients.

There are broad, regional patterns of farming systems. For example, small farms (<20 hectares), which tend to be more diverse than large farms, produce more than 75% of most foods and 80% of essential nutrients in sub-Saharan Africa, Southeast Asia, South Asia, China, and the rest of the East Asia and Pacific region.[7] Where large farms dominate the landscape, they produce the majority of the region's cereals and livestock. Globally, large farms produce more than half of the world's sugar and oil crops.[7]

Spatially, the patterns of field size show very large farms concentrated in North America, southern Australia, eastern Europe. and western South America. Small and very small farms dominate India, China, and other Asian nations.[8]

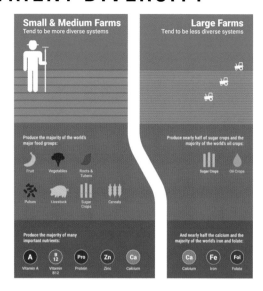

Both large and small farms are critical for food security.[7]

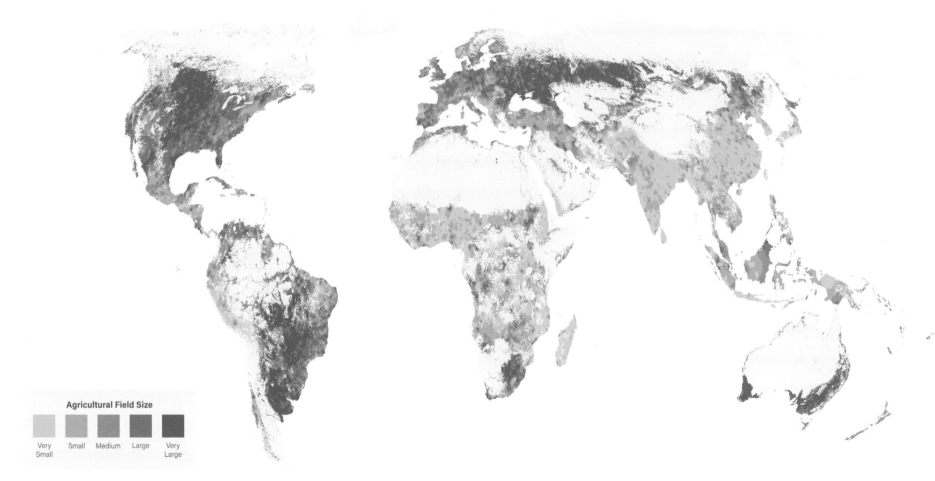

Agricultural field size was mapped[8] using a data fusion approach, combining land cover maps derived from satellite imagery, national agricultural production statistics, and crowd-sourced data from Geo-Wiki. This dataset, developed by the International Institute for Applied Systems Analysis, the International Food Policy Research Institute, and several other organizations, illustrates how multiple data sources can be combined to address the difficulty of mapping agricultural lands in many parts of the world.

Farm size around the world

As published in the Diverse Farm, Diverse Foods StoryMap.[9]

The scale of all 6 images on this spread
is the same to invite comparison:
About 1.5 miles across.

Very small farms (< 2 ha) produce more than half of most foods and nutrients in China. This includes 71% of Vitamin A and 63% of Vitamin B12. While rice is a staple crop for most Chinese farmers, relatively small field sizes allow for a higher diversity of agricultural production between fields.

Most of sub-Saharan Africa is dominated by smallholder farming, though in many cases systems are less dense than in Asia, as farms use less-productive or more arid land for grazing livestock. Small farms are responsible for over 80% of essential nutrients, and 60% of regional food calories.

Large farms account for more than 85% of protein, iron, and folate in North America and at least 77% of all other nutrients. They also account for at least three-quarters of all food, including cereal, livestock, and fruit production.

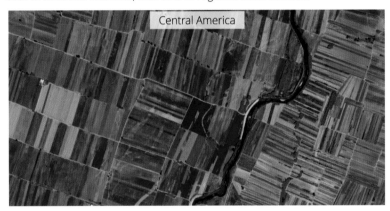

Central America, along with western Asia, North Africa, and Europe, is notable in that medium-sized farms (20–50 ha) are more significant producers. In Central America, medium-sized farms are responsible for about a quarter of most key foods, and more than 20% of all nutrient production.

This image, from the Matopiba region of Brazil, shows the massive extent of soybean farming that has become the norm in recent decades. Very large farms (>200 ha) account for more than half of all food and nutrients produced in South America, including 75% of sugar crops and oil crops.

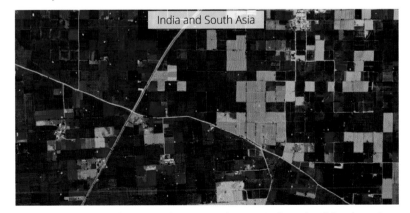

Very small and small farms produce more than 90% of nearly all foods and nutrients in India. Large farms produce less than 4% of all major nutrients. These farms are managed by millions of smallholder farmers and are vital to local food security.

TAKING STOCK OF TODAY'S FOOD PRODUCTION

Redesigning a sustainable food system requires baseline data for the basics of how much of what is produced and where: crop distribution and yields, livestock distribution and density, and cropland and pasture area. Croplands and rangelands have been mapped in several land cover data products derived from satellite imagery, such as from Landsat and MODIS.[8,10,11,12] Crop-specific distribution maps have also been developed for a few commodity crops, such as soybean and maize, for a few specific regions,[13,14] but estimating crop yields from satellite data largely remains elusive. The UN Food and Agriculture Organization (FAO) tracks national-level data on more than 150 crops. While each of these datasets has its own strengths, they do not allow for subnational analysis for crop production. To fill this information gap,

coarser-scale global maps of crop distribution and production can be created by integrating land cover map products with tabular data on yield and harvested area from agricultural census and survey statistics from counties, states, and countries.[15,16] Similar approaches of combining satellite and census data have been used to map livestock production around the world.[17,18] These tabular data on crop and livestock production can then be combined with other spatial datasets to map detailed aspects of the food system, such as where micronutrients are produced, field size, resource use, and the impact of the global food trade system. A path to sustainably improving both global food security and the environment requires assessing how the food is produced and used.

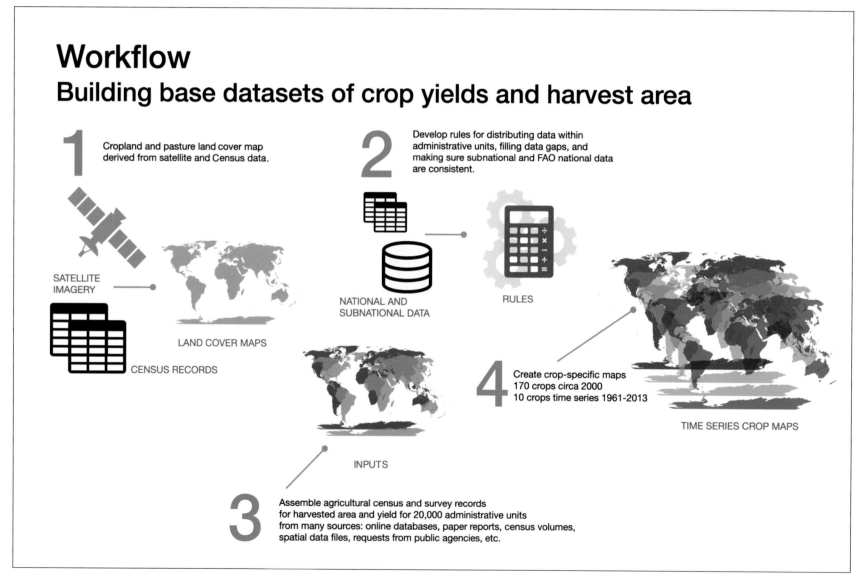

Workflow
Building base datasets of crop yields and harvest area

1 Cropland and pasture land cover map derived from satellite and Census data.

SATELLITE IMAGERY

LAND COVER MAPS

CENSUS RECORDS

2 Develop rules for distributing data within administrative units, filling data gaps, and making sure subnational and FAO national data are consistent.

NATIONAL AND SUBNATIONAL DATA

RULES

4 Create crop-specific maps
170 crops circa 2000
10 crops time series 1961-2013

TIME SERIES CROP MAPS

INPUTS

3 Assemble agricultural census and survey records for harvested area and yield for 20,000 administrative units from many sources: online databases, paper reports, census volumes, spatial data files, requests from public agencies, etc.

Agricultural census records report what is grown, but not exactly where. Satellite data reveal where crops are cultivated, but not what is grown. Once combined, these two sources of information create maps of harvested yield and area for more than 170 crops tracked by the UN Food and Agriculture Organization.

GROWING MORE FOOD ON EXISTING AGRICULTURAL LAND

About land, Mark Twain once said that "they're not making any more of it." Much of Earth's most fertile lands are already used for food production. To avoid the major environmental costs of agriculture expansion—habitat loss, water quality—more intense management on existing farmland is viewed as a critical strategy for boosting production. The research shows that yield trends are generally increasing most where wealthier countries commonly use crops for feed and fuel instead of food. But increasing corn yields in the midwestern United States, which is largely used for animal feed and fuel, does little to improve food security in countries where hunger is prevalent. Instead of aiming for marginal increases in areas already near their maximum yield, what if farmers increased yields in the lowest-performing areas? Given current crop varieties and management practices, increasing the yield for the top 16 crops to 50% of what's attainable would add enough additional calories to feed 850 million people.[19]

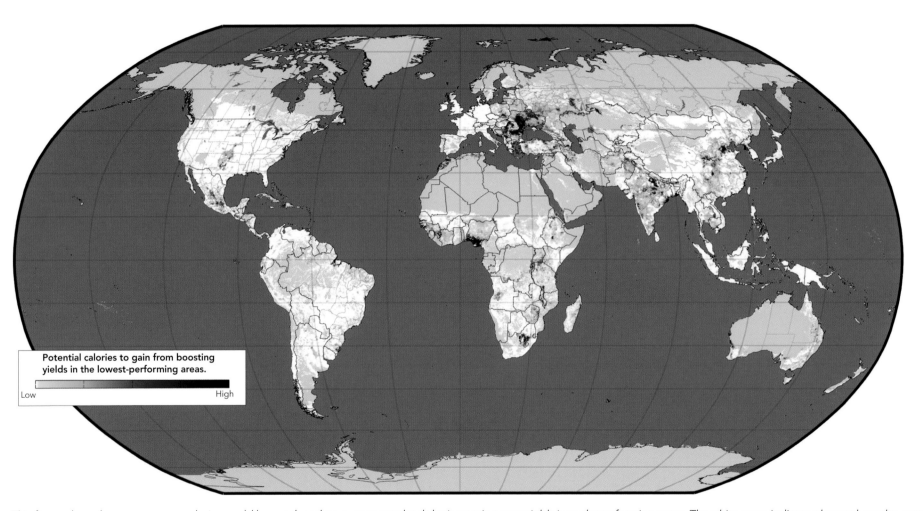

Potential calories to gain from boosting yields in the lowest-performing areas.

Low — High

This figure shows how many more calories could be produced on current croplands by increasing crop yields in underperforming areas. The white areas indicate places where the crop yields are already at least 50% of what's possible using today's best management practices.[19] The green areas represent how many additional calories could be produced by boosting yields in underperforming areas.

GROWING FOOD MORE SUSTAINABLY

Food production arguably is the biggest driver of global change. Agricultural land and production methods are major (and sometimes the main) sources of habitat loss, water use, greenhouse gas emissions, and degraded water quality. Geospatial data and analysis are critical for quantifying and mapping the status, trends, and hot spots of the environmental impacts of agricultural production. Researchers can use insights gained from analysis to assess risk, develop strategies, and track progress toward creating a sustainable food system.

Habitat loss

Agriculture is expanding most rapidly in the tropics, leading to significant habitat loss. Although deforestation rates have dramatically decreased since the 1990s, about 5.5 million hectares (13.6 million acres) of tropical forests were cleared each year from 2010–2015.[20] To counter this trend, many countries and companies established laws and commitments to reduce or eliminate deforestation from commodity supply chains. Spatial data are critical for tracking progress and compliance. The ArcGIS Emerging Hot Spot Analysis tool helps streamline analysis of forest cover trends in large datasets.[21] Online tools, such as Global Forest Watch, combine annual tree-cover data, near real-time deforestation alerts, commodity production, biodiversity, and other data to increase transparency and encourage compliance with laws and voluntary standards. ArcGIS can integrate these tools.

Water use and quality

Agriculture profoundly affects water quantity and quality. Irrigation enables agriculture to thrive in parts of the world where rainwater is inadequate in volume or falls at the wrong time. However, irrigation for farming accounts for 70% of all freshwater withdrawals, and can severely impact the environment in water-limited areas. Understanding and addressing the impacts of irrigation—and assuring wise stewardship of groundwater resources—relies on datasets of irrigation infrastructure. Using irrigation datasets,[22] models, and satellite data, other researchers estimate that about 71% of all irrigated areas have periodic water shortages[23] and that irrigation depletes aquifers in many parts of the world.[24] Globally, nearly all water used for irrigation in water-limited areas is for wheat, rice, maize, cotton, and sugarcane.[19]

Agriculture is also a major source of degraded water quality. Sedimentation and nutrients (particularly nitrogen and phosphorus) from farmlands reduce the quality of nearby streams and lakes and downstream coastal areas. The added nutrients enrich the waters, causing some species, such as algae, to thrive to the point of limiting oxygen available for fish and other animals. To identify where these sources of excess nutrients are around the world, fertilizer and manure application data can be constructed by building off the base data described earlier. Average application rates for nitrogen and phosphorus fertilizer reported elsewhere can be mapped onto the crop distribution data. Nutrient inputs from manure can be mapped using livestock density data and developing a set of rules for how the manure is distributed across pastures and crops.

Additional nitrogen inputs also come from nitrogen fixation by leguminous crops and atmospheric deposition. The nitrogen and phosphorus removed from the land is calculated as the amount of those nutrients that are in the harvested crops. From there, the amount of excess or deficit is calculated using a simple mass balance model of the inputs (fertilizer, manure, nitrogen fixation, atmospheric deposition) minus the outputs (nitrogen or phosphorus in the harvested crop).

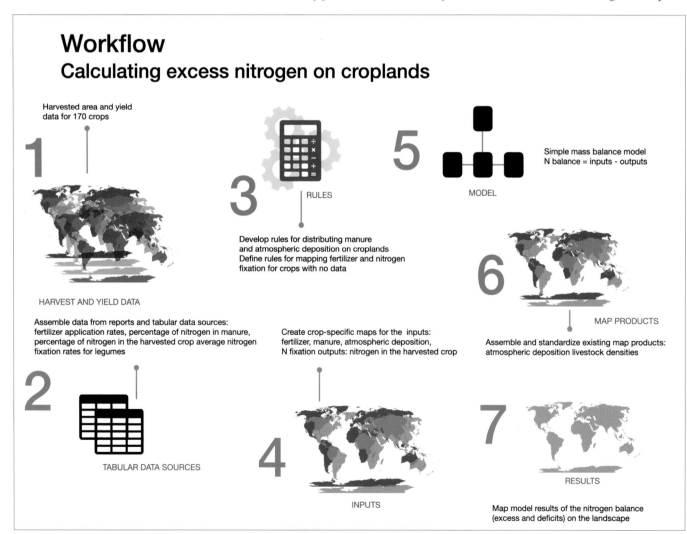

Workflow
Calculating excess nitrogen on croplands

1 Harvested area and yield data for 170 crops

HARVEST AND YIELD DATA

2 Assemble data from reports and tabular data sources: fertilizer application rates, percentage of nitrogen in manure, percentage of nitrogen in the harvested crop average nitrogen fixation rates for legumes

TABULAR DATA SOURCES

3 RULES

Develop rules for distributing manure and atmospheric deposition on croplands
Define rules for mapping fertilizer and nitrogen fixation for crops with no data

4 INPUTS

Create crop-specific maps for the inputs: fertilizer, manure, atmospheric deposition, N fixation outputs: nitrogen in the harvested crop

5 MODEL

Simple mass balance model
N balance = inputs - outputs

6 MAP PRODUCTS

Assemble and standardize existing map products: atmospheric deposition livestock densities

7 RESULTS

Map model results of the nitrogen balance (excess and deficits) on the landscape

USING WHAT WE GROW MORE EFFICIENTLY

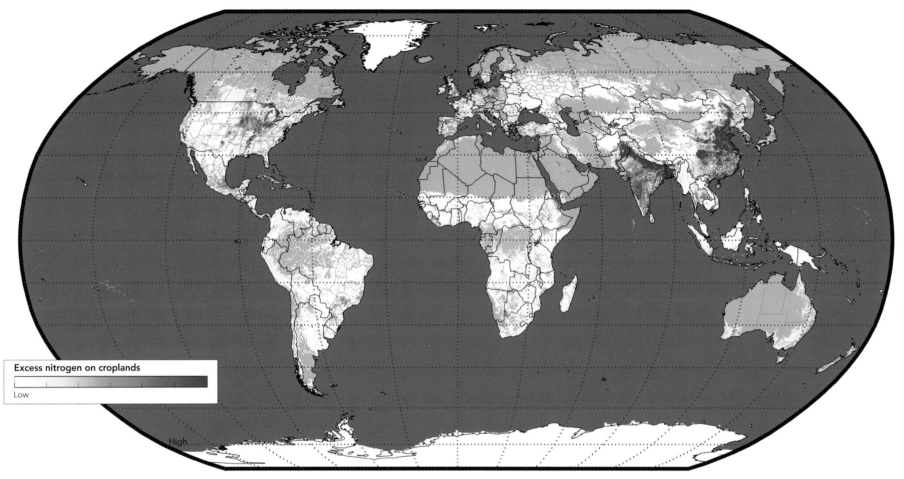

Excess nitrogen on croplands

Low — High

Excess nitrogen on croplands. Globally, about 60% of the nitrogen inputs (fertilizer, manure, nitrogen fixation, and atmospheric deposition) are in excess of the nitrogen in the harvested crop.[19] Much of this excess nitrogen leaches through the soil, into rivers, and eventually coastal areas, where high concentrations of nitrogen have created many oxygen-free "dead zones." Planners can use these assessments to target policy and management interventions to increase efficiency and reduce pollution.

Greenhouse gas emissions

Agriculture accounts for about a quarter of global greenhouse gas emissions.[5] About 80% of these emissions occur where the food is produced, with the remaining 20% from energy for producing fertilizer, processing, and transporting food.[5] The main sources of emissions associated with food production are methane from cattle and rice, nitrous oxide emissions from fertilizer application, carbon dioxide emissions from draining peatlands,[25] and land clearing. Greenhouse gas emissions associated with various farm management techniques are known, and thus by combining management information (e.g., flooding of rice fields, application of nitrogen fertilizer, draining peatlands) with maps of crop production, we can identify hot spots of emissions.

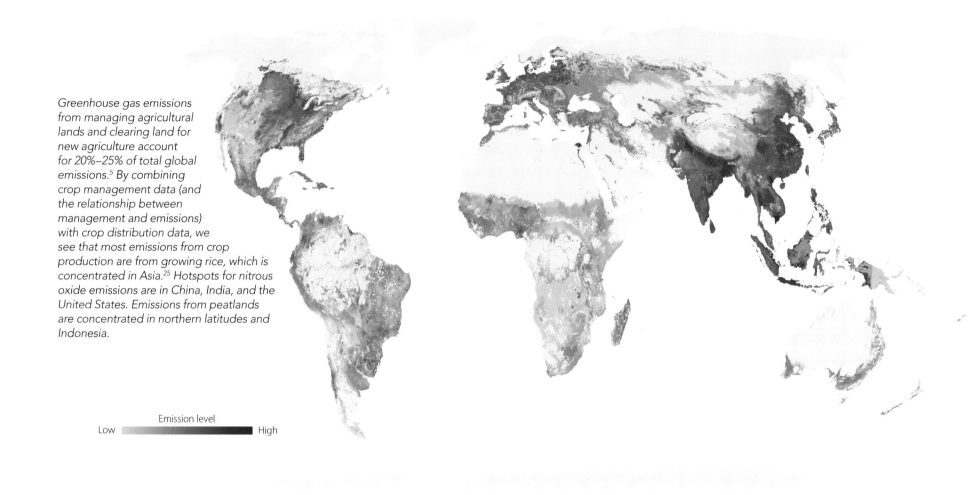

Greenhouse gas emissions from managing agricultural lands and clearing land for new agriculture account for 20%–25% of total global emissions.[5] By combining crop management data (and the relationship between management and emissions) with crop distribution data, we see that most emissions from crop production are from growing rice, which is concentrated in Asia.[25] Hotspots for nitrous oxide emissions are in China, India, and the United States. Emissions from peatlands are concentrated in northern latitudes and Indonesia.

Emission level

Low ▬▬▬▬▬▬ High

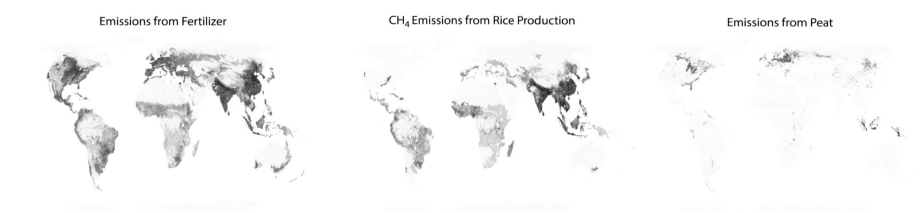

Emissions from Fertilizer   CH₄ Emissions from Rice Production   Emissions from Peat

FOOD VERSUS FEED

Improving food security is not just about producing more food. Changing what we grow, what we eat, and what we waste can have a greater impact on food security and the environment. About 30–50% of food is wasted.[26] Similarly, 36% of calories produced on the world's croplands are used for animal feed, and many of those calories are "lost" in the food system, as it takes several calories of feed to produce a calorie of meat or dairy.[27] Reducing food waste and eating less meat are two actions that not only reduce the need for more food, but also the land, water, and other natural resources used to produce it.

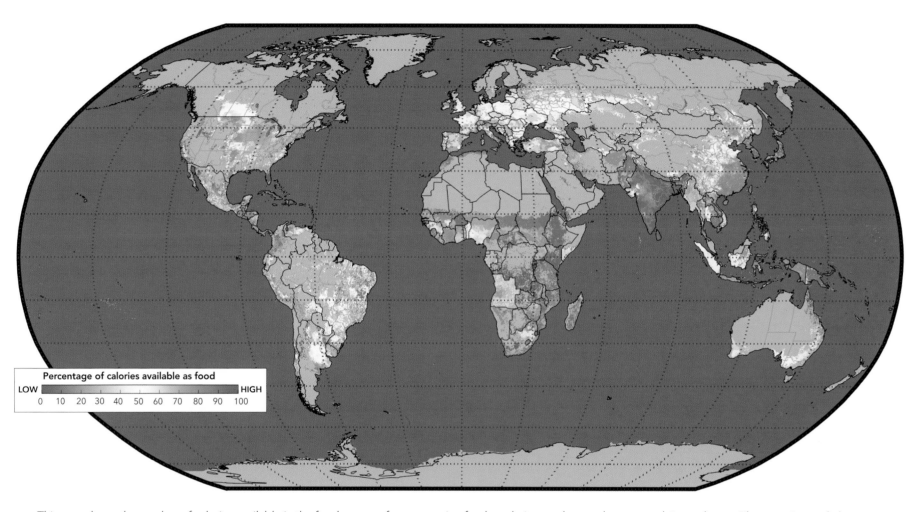

Percentage of calories available as food

LOW ——————————————— HIGH
0 10 20 30 40 50 60 70 80 90 100

This map shows the number of calories available in the food system after accounting for the calories used to produce meat, dairy, and eggs. The areas in purple have few calories available that end up in the food system, whereas the green areas are places where calories produced are consumed directly.[27]

CASE STUDY

Targeting efforts to create regenerative ranching and agricultural systems across Latin America

Researchers and others are using global datasets to identify opportunities for creating sustainable agricultural systems across Latin America. Farmers use nearly 38% of the region's land for agriculture.[28] Latin America raises 28% of the world's cattle,[29] and 47% of global soybean exports originate in Brazil and Argentina alone.[29] Many of these economic strengths resulted in habitat loss in several biodiversity hot spots within the region. Further, land degradation and a changing climate create risks for producing food and conserving biodiversity.

The Nature Conservancy (TNC) and other conservation nonprofits increasingly focus their efforts on sustainable agriculture to benefit people's livelihoods and biodiversity. Using a geodesign process, TNC, the University of Minnesota's Institute on the Environment, the International Center for Tropical Agriculture, and the University of São Paulo collaborated to identify a set of "action landscapes." Their efforts focused on these landscapes to maintain and restore regenerative ranching and agricultural systems that increase profit, improve the environment, and build

resilience to climate change. The collaboration prepared several iterations of analysis and maps to create a set of products that could support the data-driven decision-making. The products aimed to provide context for people with on-the-ground expertise and help identify major opportunities and challenges across the region. The partnership created map products for several attributes of TNC's Regenerative Ranching and Agriculture strategy: agricultural productivity, climate-related risks, degraded lands, and restoration of ecosystem services. Participants reviewed map products, developed selection criteria, and identified a draft set of priority landscapes based on visual interpretation of maps. Later, they completed additional analysis and revised maps to refine the products and ensure that the data supported the draft set of priorities as well as to include additional landscapes that met all criteria.

TNC uses this product in Latin America to guide investments in these landscapes and design actions with partners.

Learning from farmers helps shape regional strategies. Here, a ranching family in Colombia shares its three generations of experience restoring degraded land to sustainably increase meat and milk production, profits, biodiversity, and soil health. In other cases, collaborations with large-scale farmers producing commodities like soy help determine which management practices effectively increase production, build climate resilience, and reduce environmental impacts.

The planning process identified a set of action landscapes for The Nature Conservancy and its partners to focus their efforts to create regenerative ranching and agricultural systems. The landscape boundaries shown here are being modified as implementation proceeds.[30]

CONCLUSION

GIS is an essential tool for creating a sustainable food system. The examples in this chapter show how spatial data and analysis are used to advance three broad solutions to grow more food on existing agricultural lands, produce food more sustainably, and use what we already produce more efficiently. These products help instituions—including development banks, nonprofit organizations, governments, philanthropic foundations, and companies—develop science-based strategies and target their investments.

NOTES

Photo on opening spread by Nena Terrell/USAID.

1. Tilman, D., Balzer, C., Hill, J. & Befort, B. L. "Global Food Demand and the Sustainable Intensification of Agriculture." *Proceedings of the National Academy of Sciences.* 108, 20260–20264 (2011).

2. Ray, D.K., Mueller, N.D., West, P.C., and Foley, J.A. "Yield Trends Are Insufficient to Double Global Crop Production by 2050." *Public Library of Science ONE [PLOS].* doi: 10.1371/journal.pone.006642 (2103).

3. Foley, J. A. et al. "Solutions for a Cultivated Planet." *Nature* 478, 337–342 (2011).

4. Gleick, P. H., Cooley, H. & Morikawa, M. *The World's Water 2008-2009: The Biennial Report on Freshwater Resources.* (Island Press, 2009).

5. Vermeulen, S. J., Campbell, B. M. & Ingram, J. S. I. "Climate Change and Food Systems." *Annual Review of Environmental Resources.* 37, 195–222 (2012).

6. Ramankutty, N., Evan, A., Monfreda, C., and Foley, J.A. "Farming the Planet 1: Geographic Distribution of Global Agricultural Lands in the Year 2000." *Global Biogeochemical Cycles.* 22, GB1003 (2008).

7. Herrero, M., et al. "Farming and the Geography of Nutrient Production for Human Use: A Transdisciplinary Analysis." *The Lancet Planetary Health.* 2017.

8. Fritz, S. et al. "Mapping Global Cropland and Field Size." *Global Change Biology* 21, 1980–1992 (2015).

9. *Diverse Farms, Diverse Foods.* Available at: https://umn.maps.arcgis.com/apps/Cascade/index.html?appid=a48c26df4577490ba8b92d410df2e1fd.

10. ESA - Climate Change Initiative. *Global Land Cover.* (2016). Available at: https://www.esa-landcover-cci.org/?q=node/175. (Accessed: January 25, 2019).

11. Teluguntla, P. G. et al. *Global Cropland Area Database* (GCAD) derived from "Remote Sensing in Support of Food Security in the Twenty-First Century: Current Achievements and Future Possibilities." In *Land Resources: Monitoring, Modelling, and Mapping* (Taylor & Francis, 2015).

12. Friedl, M. A. et al. "MODIS Collection 5 Global Land Cover: Algorithm Refinements and Characterization of New Datasets." *Remote Sensing of Environment* 114, 168–182 (2010).

13. Agrosatélite Applied Geotechnology. Soya Area. (2018). Available at: http://biomas.agrosatelite.com.br/#/index. (Accessed: January 25, 2019).

14. Jin, Z. et al. "Smallholder Maize Area and Yield Mapping at National Scales with Google Earth Engine." *Remote Sensing of Environment* 228, 115–128 (2019).

15. Monfreda, C., Ramankutty, N. & Foley, J. "Farming the Planet: 2. Geographic Distribution of Crop Areas, Yields, Physiological Types, and Net Primary Production in the Year 2000." Global Biogeochem. *Cycles* 22, GB1022 (2008).

16. Ray, D. K. et al. "Climate Change Has Likely Already Affected Global Food Production." *PLoS One* 14, e0217148 (2019).

17. Herrero, M. & Thornton, P. K. "Livestock and Global Change: Emerging Issues for Sustainable Food Systems." *Proceedings of the National Academy of Sciences of the USA* 110, 20878–20881 (2013).

18. Gilbert, M. et al. "Global Distribution Data for Cattle, Buffaloes, Horses, Sheep, Goats, Pigs, Chickens, and Ducks in 2010." *Scientific Data* 5, 180227 (2018).

19. West, P. C. et al. "Leverage Points for Improving Global Food Security and the Environment." *Science* 345, 325–328 (2014).

20. Keenan, R. J. et al. "Dynamics of Global Forest Area: Results from the FAO Global Forest Resources Assessment 2015." *Forest Ecological Management* 352, 9–20 (2015).

21. Harris, N. L. et al. "Using Spatial Statistics to Identify Emerging Hot Spots of Forest Loss." *Environmental Research Letters* 12, 024012 (2017).

22. Siebert, S. et al. "A Global Data Set of the Extent of Irrigated Land from 1900 to 2005." *Hydrology and Earth Systems Sciences* 19, 1521–1545 (2015).

23. Brauman, K. A., Richter, B. D., Postel, S., Malsy, M. & Flörke, M. "Water depletion: An Improved Metric for Incorporating Seasonal and Dry-Year Water Scarcity into Water Risk Assessments." *Elementa: Science of the Anthropocene* 4, 000083 (2016).

24. Gleeson, T., Wada, Y., Bierkens, M. F. P. & van Beek, L. P. H. "Water Balance of Global Aquifers Revealed by Groundwater Footprint." *Nature* 488, 197–200 (2012).

25. Carlson, K. M. et al. "Greenhouse Gas Emissions Intensity of Global Croplands." *Nature Climate Change* 7, 53–78 (2016).

26. Gustavsson, J., Cederberg, C., Sonesson, U., van Otterdijk, R., and Meybeck. A. "Global Food Losses and Food Waste: Extent, Causes and Prevention." *Rome*, 2011.

27. Cassidy, E., West, P., Gerber, J., and Foley, J. 2013. "Redefining Agricultural Yields: From Tonnes to People Nourished per Hectare." *Environmental Research Letters* 8:034015.

28. World Bank. World Bank Open Data. (2018). Available at: https://data.worldbank.org/indicator/AG.LND.AGRI.ZS?locations=ZJ. (Accessed: 10th January 2019).

29. United Nations Food and Agriculture Organization.

30. *See Action Landscapes.* Available at: https://www.arcgis.com/apps/Cascade/index.html?appid=dd7bd5aeb9344725a4f9d852e4056d9e.

Crop sprayer photo by Kevin Casper.
Ranching family photo by Paul West.

TRACKING GLOBAL FOREST LOSS

Supplied with big geodata, researchers at the World Resources Institute are using advanced geospatial tools and data frameworks to better monitor and model the spatial patterns of human activity in the world's remaining tropical forest landscapes.

By Elizabeth Goldman and Nancy Harris, **WRI**; Lauren Bennett, **Esri**; and Stephen Ansari, Christopher Gabris, and Michael Lippmann, **Blue Raster**

The edge of deforestation in the Democratic Republic of the Congo. The biggest drivers of deforestation in the Congo rainforest during the past 20 years have been small-scale subsistence agriculture, clearing for charcoal and fuel wood, urban expansion, and mining. Industrial logging has been the biggest driver of forest degradation. Image by Axel Fassio/CIFOR.

THE COMPETITION FOR FORESTS

Planetwide, forests have changed rapidly during the past several decades. Rising global demand for commodities such as timber, wood fiber, palm oil, beef, and soy has pushed agricultural land into previously forested areas. The forests reflect changes in national political and economic conditions. Forests provide wood for construction, fiber for paper, and fuel for energy. They offer food, medicines, and other nontimber forest products. They moderate the quality, quantity, and timing of freshwater flows and influence regional precipitation patterns (critical for nearby agriculture and cities). Forests are central to the fight against climate change because they remove carbon from the air and emit it when burned, cleared, or degraded. They offer a place for recreation and spiritual renewal and are home to 70 million Indigenous peoples. And forests harbor the most biodiversity of any ecosystem on Earth.

But the forests, and their capacity to provide these benefits, are threatened. About 10,000 years ago in the age before humans learned to farm, forests covered about half of all land on Earth. Approximately half of these forests have since been cleared. Most forests still standing today have been degraded or fragmented; by one measure, less than one-fifth of them are still intact. The main causes of forest loss and degradation include the expansion of agriculture and settlements, unsustainable extraction of timber and fuel wood, and roads and other infrastructure that fragment forests and bring settlers to new frontiers. Climate change impacts, including severe fires and new vectors and outbreaks of forest pests and diseases, exacerbate the decline.

Underlying causes of forest loss and degradation include market and governance failures driving land-use choices that do not recognize the value of forests or mitigate the risks of depleting them. For instance, decisions about the fate of a forest are often made in the absence of accurate information, in a nontransparent manner, without participation of all relevant stakeholders, and without adequate accountability. In some places, corruption and powerful vested interests hold sway, governance is opaque, and laws are poorly enforced. And poor recognition of customary rights to forest lands and resources fans conflict and robs people of their cultural heritage and livelihoods.

During the past decade, many governments and companies have made time-bound commitments to end deforestation, restore degraded forest landscapes, and achieve sustainable forest management. The UN Sustainable Development Goals, the New York Declaration on Forests (NYDF), the Paris Climate Agreement, and the Bonn Challenge provide policy frameworks, accountability mechanisms, and financing opportunities to help these goals succeed. These global and local policies such as the moratoria on agricultural development in Brazil and Indonesia, have been enacted with much fanfare. Policy makers and local enforcement agencies all need timely, reliable, and trustworthy data to track individual and collective progress, guide decisions about where and how to invest, and inform the design and implementation of policies and programs.

Instituted by the World Resources Institute, Global Forest Watch is an online platform that synthesizes data from authoritative sources and provides geodata and tools for monitoring forests, providing near real-time information about where and how forests are changing around the world.

A landscape under change, this region near Yangambi in Democratic Republic of the Congo shows the effects of agricultural encroachment. Image by CIFOR.

GLOBAL FOREST WATCH

Launched in 2014, the Global Forest Watch (GFW) platform provides timely and spatially detailed information on forest dynamics that is globally consistent and locally relevant. The GFW platform enhances the practical use of these data by providing solutions to the challenges often associated with big data, including visualization, storage, analysis, sharing, and querying. These and other data products derived from satellite imagery have fundamentally changed the way the world's forests are monitored by various stakeholders. But as sources of data become larger, more complex, and more numerous, the ability to quickly explore and interpret patterns with confidence has become a bottleneck for effectively using these data to inform forest policy and management decisions.

Products such as the University of Maryland's 30-meter-resolution GLAD (Global Land Analysis and Discovery) laboratory forest-loss data and weekly deforestation alerts (available on globalforestwatch.org) have made near real-time forest monitoring a reality. Forest managers, law enforcement officials, and policy makers can use this information to track how forests fare and identify deforestation while there's still time to make a difference.

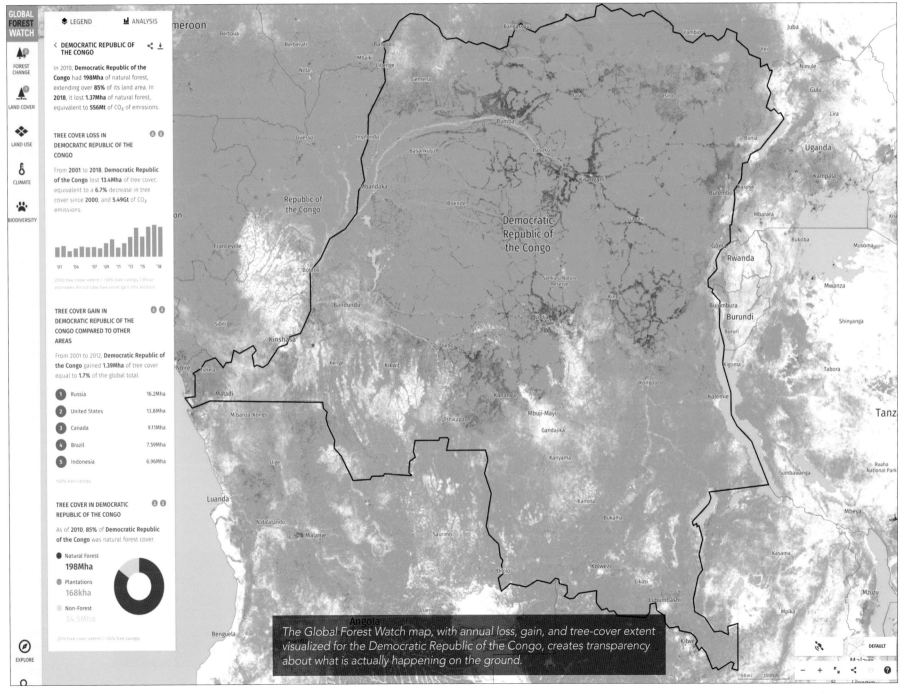

The Global Forest Watch map, with annual loss, gain, and tree-cover extent visualized for the Democratic Republic of the Congo, creates transparency about what is actually happening on the ground.

SPATIOTEMPORAL CLUSTERS OF FOREST LOSS

To turn this wealth of forest data into actionable information, researchers needed to apply methods that would illuminate the spatial and temporal trends that exist in the data. The goal was to find meaningful areas of primary forest loss across the tropics based on spatial statistics and the latest analytical methods rather than easily biased interpretations of thematic maps. Geospatial appliations often use the term *hot spot* to describe a region or value that is higher relative to its surroundings. In the context of forest conservation, deforestation hot spots can be thought of as *fronts*—broad regions of deforestation concern based on expert opinion and scenario analyses where available.

In GIS methodological terms, a hot spot is defined as an area that exhibits statistically significant clustering in the spatial pattern of forest loss. Hot spots are locations where observed patterns are not likely the result of random processes or of subjective cartographic design decisions; they represent places where underlying spatial processes are at work. Emerging hot spot analysis extends this definition to incorporate information about the temporal dimension of the data. The Emerging Hot Spot analysis tool, which is part of the Space Time Pattern Mining toolbox in ArcGIS® Pro, allows researchers to understand spatial clusters of deforestation across the tropics and trends in those clusters over time. Results show new, sporadic, persistent, intensifying, and diminishing hot spot patterns, as seen in the table.

In the graphic showing the workflow for emerging hot spots, the task is revealed as a high-intensity geospatial operation. Cell-based raster data derived from raw satellite imagery are aggregated and clipped into a space-time cube (essentially a set of georeferenced layers stacked in time slices). An analysis mask is applied to include only relevant, forested areas.

These spatiotemporal patterns are based on statistical analysis of the clustering of instances of primary forest loss, where statistically significant hot spots of primary forest loss represent places where researchers found more clustering of primary forest loss than would be expected based on random spatial processes. Finally, statistical results were summarized into categories as shown in the table to help users of the data interpret and communicate the information.

Hot spot definitions

Time period	Hot Spot Type	Definition
Short term trend (Less than 16 years were statistically significant)	New	A hotspot for only the year 2018. It has never been a hot spot before.
	Sporadic	An on-again then off-again hot spot. Less than 16 years have been hot spots.
Long term trend (At least 16 years were statistically significant)	Persistent	A statistically significant hot spot for 16 years with no discernible increasing or decreasing trend in the intensity of loss over time.
	Intensifying	A statistically significant hot spot for 16 years, including the year 2018. In addition, the intensity of clustering of high counts in each year is increasing overall and that increase is statistically significant.
	Diminishing	A statistically significant hot spot for 16 years, including the year 2018. In addition, the intensity of clustering of high counts in each year is decreasing overall and that decrease is statistically significant.

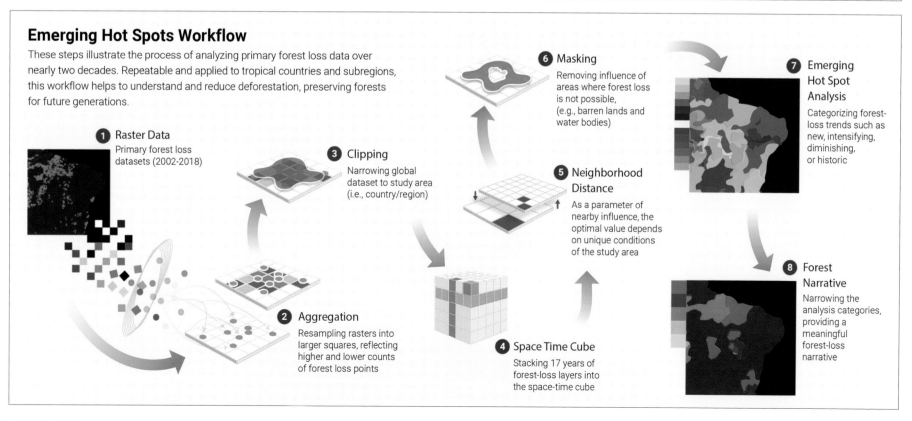

Emerging Hot Spots Workflow

These steps illustrate the process of analyzing primary forest loss data over nearly two decades. Repeatable and applied to tropical countries and subregions, this workflow helps to understand and reduce deforestation, preserving forests for future generations.

1 Raster Data
Primary forest loss datasets (2002-2018)

2 Aggregation
Resampling rasters into larger squares, reflecting higher and lower counts of forest loss points

3 Clipping
Narrowing global dataset to study area (i.e., country/region)

4 Space Time Cube
Stacking 17 years of forest-loss layers into the space-time cube

5 Neighborhood Distance
As a parameter of nearby influence, the optimal value depends on unique conditions of the study area

6 Masking
Removing influence of areas where forest loss is not possible, (e.g., barren lands and water bodies)

7 Emerging Hot Spot Analysis
Categorizing forest-loss trends such as new, intensifying, diminishing, or historic

8 Forest Narrative
Narrowing the analysis categories, providing a meaningful forest-loss narrative

EXPLORING THE RESULTS: BRAZIL

Emerging hot spot analysis applied to the University of Maryland primary forest-loss dataset revealed important trends in loss from the years 2002–2018. Old growth, or *primary*, tropical rain forests are a crucially important forest ecosystem, containing trees that can be hundreds or even thousands of years old. They store more carbon than other forests and are irreplaceable in sustaining biodiversity. Primary rain forests provide habitat for animals ranging from orangutans and mountain gorillas to jaguars and tigers. Once cut down, these forests may never return to their original state. The tropics-wide analysis examined results from four countries that contain some of the most important forests in the world, where the impact of environmental, economic, and political changes on these forests must be better understood. For the latest analysis and primary forest-loss data updates, visit globalforestwatch.org.

From 2002 to 2018, Brazil lost an average of 1.4 million hectares of primary forest per year, an area about the size of the Bahamas. The government enacted policies such as the Amazon Soy Moratorium in the early 2000s to curb deforestation in the Amazon rainforest. Nevertheless, Brazil experienced its third-highest rate of primary forest loss in 2018 after a prominent fire-related spike in 2016–2017.

In this analysis, it is still too early to assess how the devastating fires from the summer of 2019 and the weakening of environmental laws and enforcement under Brazil's Bolsonaro administration will impact forest loss. The high rate of primary forest loss in 2018 occurred before President Jair Bolsonaro took office (though there is evidence of deforestation rates spiking during the election season).

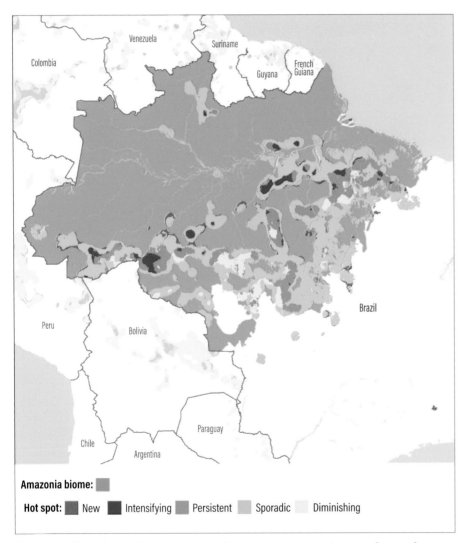

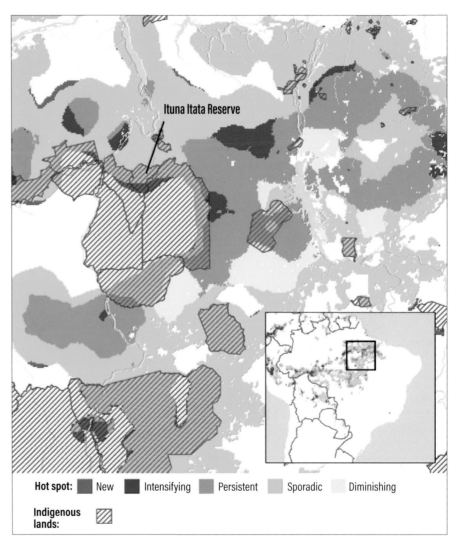

The maps of Brazil show hot spots across the country, representing new fronts of primary forest loss. Wildfires caused some of the 2018 loss, but most of it can be attributed to clear-cutting in the Amazon, threatening to reverse the declines in deforestation the country achieved in the early 2000s.

Notably, several hot spots of primary forest loss occurred near and within indigenous territories. For example, the Ituna Itata Reserve had more than 4,000 hectares illegally cleared in the first half of 2018, more than double the total loss from 2002–2017. The reserve is home to some of the world's last remaining uncontacted peoples, who depend on the forest for survival and have conserved it for centuries.

Colombia

Primary forest-loss data for Colombia reflects a 9 percent increase in primary forest loss between 2017 and 2018, continuing a dramatic upward trend that began in 2016. Three regions on the border of the Amazon biome (Meta, Guaviare, and Caquetá) show new hot spots of loss advancing into pristine intact forest landscape, and these areas account for about half of the increase that occurred in Colombia in 2018.

The rapid increase in forest loss happened as peace came to the country. In 2016, the Revolutionary Armed Forces of Colombia (FARC), the country's largest rebel group, was pushed out of large amounts of remote forest it previously controlled. The FARC had kept tight control over land use and allowed little commercial use of resources. After FARC demobilized, a power vacuum emerged, leading to illegal clearing for pasture and coca, mining, and logging by other armed groups.

Land speculation is rampant, as people occupy and deforest new areas in the hopes of getting a land title under the rural reform law, a key component of the peace agreement. Abandoned FARC trails also provide access to previously remote forest areas, with some regional governments officially expanding these roads to promote development. New hot spots in the northeastern part of the country could also indicate loss associated with some migration across the border from Venezuela.

The Colombian government is actively working to slow forest destruction. It canceled a major highway project connecting Venezuela and Ecuador, destroyed several illegal roads, expanded Chiribiquete National Park by 1.5 million hectares, and launched the Green Belt initiative to protect and restore a 9.2-million-hectare forest corridor. Norway, Germany, and Britain have also pledged to spend up to $366 million from now to 2025 to slow deforestation in the Colombian Amazon. It's too early to tell whether these actions and others will be enough to slow the country's rampant forest loss, but evaluating the data again with the Emerging Hot Spot tool and looking for diminishing and historical patterns will help us to better understand the rapidly changing conditions in this country.

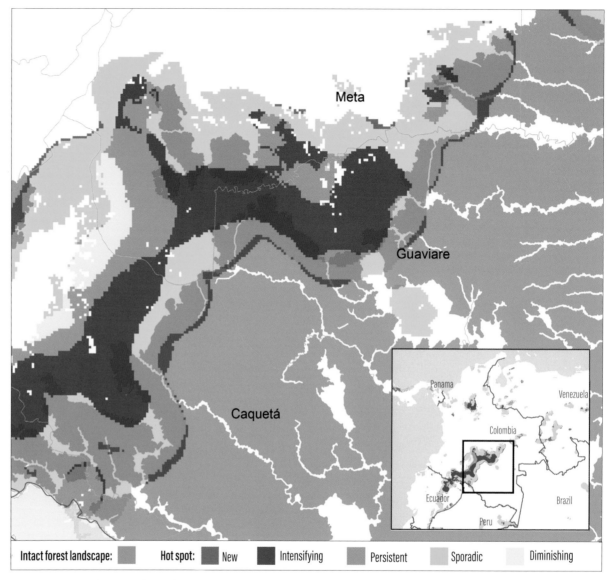

Intact forest landscape: | Hot spot: New | Intensifying | Persistent | Sporadic | Diminishing

New hot spots of primary forest loss encroach into intact forest landscapes in the Colombian states of Meta, Guaviare, and Caquetá. Intact forest landscapes are pristine forests with little human-caused deforestation.

The Democratic Republic of the Congo

The Democratic Republic of the Congo (DRC) primary forest-loss data reflect the country's conflicting environmental regulations and lack of enforcement. From 2002 to 2018, DRC lost an average of 256,000 hectares (about 632,590 acres) per year, an area about the size of Luxembourg, and reached its second-largest total loss in 2018. Agriculture, artisanal logging, and charcoal production drive much of the forest loss in the region, with nearly 75 percent of DRC forest loss in 2018 occurring in shifting cultivation areas known as the rural complex. Shifting cultivation is a type of rotational farming system, in which trees are cleared and the land is farmed for several years. Once soil nutrients can no longer support agriculture, the land is left fallow, and trees and secondary forest regrow until eventually the vegetation is cleared again for agricultural activities.

While shifting cultivation does not necessarily indicate expansion into primary forest, growing populations can intensify agricultural practices, thus reducing fallow periods during which trees regrow naturally. Overlaying the shifting cultivation areas with emerging hot spot results reveals overlapping areas with the sporadic hot spot category. This on-again, off-again category matches the cadence of periodic tree clearing that typically occurs in shifting cultivation areas.

In addition to sporadic hot spots, some shifting cultivation also falls into new and intensifying hot spot categories, as well as no hot spot categories, meaning no statistically significant pattern was detected or a statistically significant low amount of forest loss occurred. Further research could compare shifting cultivation practices and other local conditions in the expected sporadic hot spot areas, with conditions in the intensifying, new, and non hot spot areas determining whether changes in population or other agricultural practices are impacting forest loss.

While shifting cultivation is associated with much of the forest loss observed in DRC recently, changes in federal law could see pressure on forests shift to new areas. For the past 16 years, DRC has had a moratorium on new industrial logging concessions, but the government reinstated concessions to two companies in 2018. Environmentalists worry that opening the forest to additional logging could exacerbate the country's growing deforestation problem. But there is more to DRC's forest loss than industrial logging concessions. While the moratorium applied only to industrial logging, artisanal logging, often illegal, also soared. Given the increasing trends observed since 2016, it is critical that DRC move ahead with improved land-use planning and forest law enforcement, and vigorously transition to better management practices.

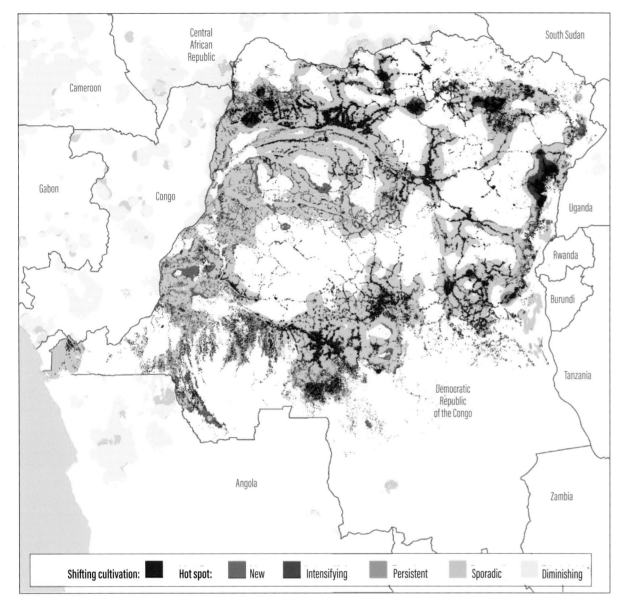

In DRC, overlaying shifting cultivation with emerging hot spots reveals overlapping areas with the sporadic hot spot category. This on-again, off-again category matches the cadence of periodic tree clearing that typically occurs in shifting cultivation areas.

Shifting cultivation: ■ Hot spot: ■ New ■ Intensifying ■ Persistent ■ Sporadic ■ Diminishing

Indonesia

Indonesia lost about 538,000 hectares (about 1,329,427 acres) of forest annually between 2002 and 2018, an area about the size of Brunei. Unlike most tropical forests, Indonesia experienced a drop in forest loss in 2017 and 2018, including a 40 percent decline in primary forest loss in 2018 compared to the average annual rate from 2002–2016. The country experienced an even more dramatic decline in forest loss in protected forests, suggesting that recent government policies are working.

On peatlands deeper than 3 meters, which have been legally protected from development since 2016, forest loss dropped 80 percent from the 2002–2016 average. And in areas under Indonesia's forest moratorium, primary forest loss dropped 45 percent in 2018 compared with 2002–2016.

Normally, the presence of large agricultural tree plantations such as oil palm complicates the task of measuring loss, especially in Sumatra, Kalimantan, and Papua. Because tree-cover loss data don't distinguish between natural vegetation and planted trees, harvest activity within plantation boundaries is observed within the tree-cover loss data. However, since the team used the primary forest extent data in the emerging hot spot analysis, this removed the expected loss within plantations and draws attention to more concerning trends occurring within primary forests.

Using the emerging hot spot analysis to dive further into the primary forest-loss trends, the research team discovered several concerning new hot spots of forest loss in protected areas. Two areas in Kerinci Seblat National Park in Sumatra overlap with new hot spots, and additional research reveals that small areas of forest have been cleared for agriculture and other purposes in this protected area.

In a country with recent forest-loss decline but also thousands of smaller, fragmented areas of loss to investigate, it can be daunting to try to understand where to focus deforestation reduction efforts. When paired with contextual layers such as protected areas, emerging hot spots can be especially helpful in evaluating the statistical importance of loss and identifying the most concerning areas of forest loss.

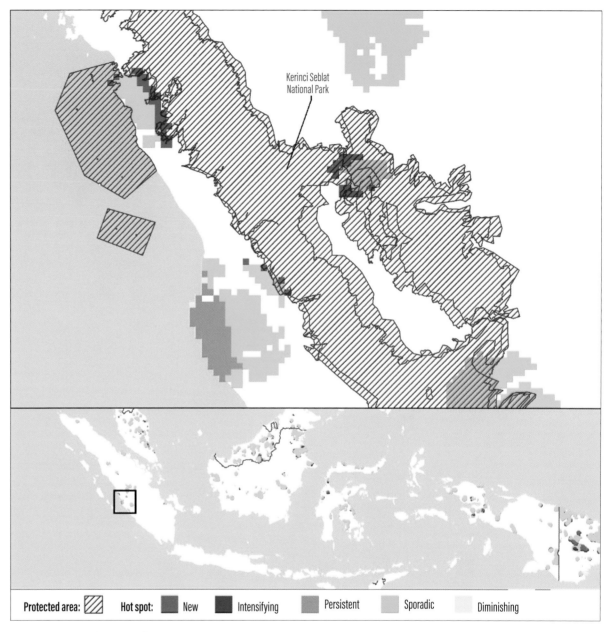

New hot spots of primary forest loss overlap with Kerinci Seblat National Park on Sumatra, Indonesia. Forests in this area have been cleared for agricultural activities.

GETTING TO EMERGING HOT SPOTS

The foundation of the analysis is the emerging hot spot analysis, which identifies spatiotemporal trends in the clustering of data, in this case, with a focus on finding clusters of forest loss. It takes as input a space-time cube, created from the forest loss raster data through a process of reclassification and aggregation. The space-time cube represents the data in a cube-like structure with information about what has happened at each location over time. The analysis calculates a spatiotemporal Getis-Ord Gi* statistic. The statistic tests to see whether there is clustering of deforestation in space and time, and, more importantly, to see whether that clustering is more than the team would expect to see based on random chance.

The team made several more important decisions to fine-tune the analysis based specifically on the exploration of forest-loss data. First, the team used a dataset representing the extent of forest in each country as a mask for the analysis. This mask limited the locations evaluated in the analysis to include only relevant locations. This was a critical aspect of the workflow, because every location included in the analysis contributed to the global average against which all locations were compared. When locations are included that either do not have forests or are not relevant to the analysis, the global average can become unrepresentative and lead to unreliable results. Additionally, while the default neighborhood size was helpful as a first iteration, it was frequently found to be too large for the forest-loss data. Usually a distance of about two-thirds the default brought out additional, nuanced patterns in the loss data. Ideally, the neighborhood size is selected based on the width of loss patterns in the landscape, such as that along roads and rivers, or of typical patchy loss, such as farms or pasture.

Once the space-time hot spot analysis completes, each bin in the input cube has an associated z-score and p-value added to it. These hot and cold spot trends are evaluated using the Mann-Kendall trend test on the z-scores at each bin over time. Those trends, along with the z-score and p-value for each bin, are then used to categorize each study area location (each location is composed of a time series of space-time bins).

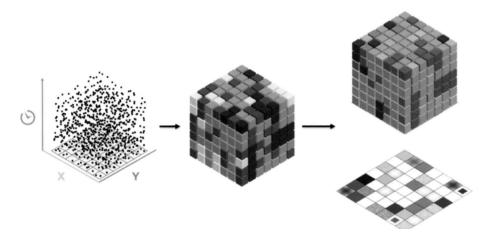

An emerging hot spot analysis starts with an aggregation process, turning the global forest loss raster data into a series of space-time cubes representing the spatiotemporal forest-loss data in space and time. Next, the statistical analysis evaluates each location to determine whether it is part of a statistically significant hot or cold spot based on a comparison of local values (in space and time) and global values. Those locations marked as hot or cold are locations where it is unlikely that the clustering observed happened as the result of random chance. Finally, those spatiotemporal hot and cold spots are summarized into categories that help communicate the trends in clustering over time.

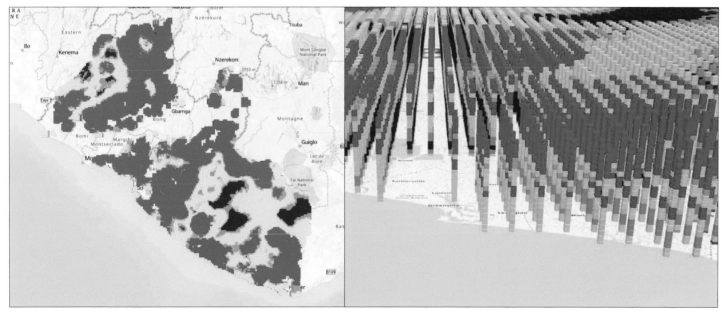

The hot and cold spots of primary forest loss in Liberia as seen in the red and blue areas, respectively, can be stacked into a 3D visualization to see the underlying patterns that ultimately make up the hot spot categories. On their own, they tell only part of the story of forest loss in Liberia. The categorization plays a critical role in the interpretation and communication of the results.

INTERPRETATION ACROSS THE TROPICS

The Emerging Hot Spot tool outputs 17 categories of results: eight hot spot types, eight cold spot types, and a no-pattern-detected category. The results identify different long- and short-term trends, with variations in the intensity of clustering over time, and the presence or absence of hot and cold spots at different points in the time series. Since the researchers mostly care about where forests are experiencing important loss, they chose to recategorize locations by eliminating all cold spots (clusters of statistically significant low amounts of forest loss) and combined and eliminated other categories to focus only on the stories that would matter most to policy makers. This resulted in five categories, which are outlined before the results section.

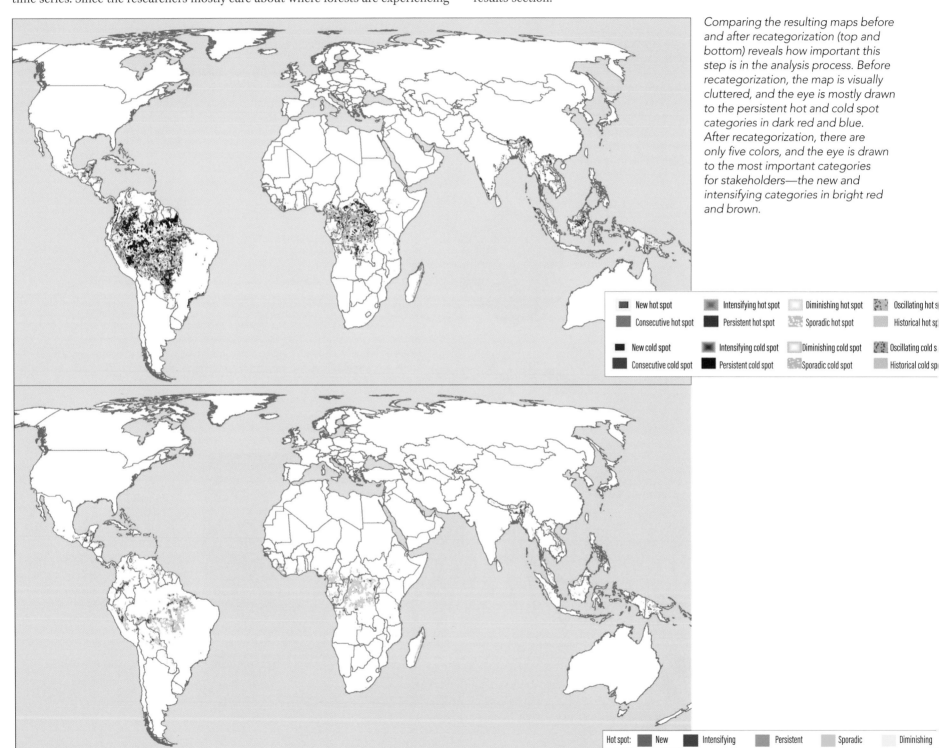

Comparing the resulting maps before and after recategorization (top and bottom) reveals how important this step is in the analysis process. Before recategorization, the map is visually cluttered, and the eye is mostly drawn to the persistent hot and cold spot categories in dark red and blue. After recategorization, there are only five colors, and the eye is drawn to the most important categories for stakeholders—the new and intensifying categories in bright red and brown.

New hot spot · Intensifying hot spot · Diminishing hot spot · Oscillating hot spot
Consecutive hot spot · Persistent hot spot · Sporadic hot spot · Historical hot spot
New cold spot · Intensifying cold spot · Diminishing cold spot · Oscillating cold spot
Consecutive cold spot · Persistent cold spot · Sporadic cold spot · Historical cold spot

Hot spot: New · Intensifying · Persistent · Sporadic · Diminishing

CONCLUSIONS

Emerging hot spot analysis can provide powerful information about precisely which places experience the most significant impacts of deforestation, a valuable insight for decision-makers, law enforcement, journalists, and activists. Focusing on the places with the largest, new clusters of clearing can help decision-makers, law enforcement, and other forest advocates apply their limited resources in a way that will have the most impact.

The emerging hot spots of primary forest-loss layer are now available on the GFW platform. Several organizations have adopted the methodology in their work to combat forest loss. The Monitoring of the Andean Amazon Project has used emerging hot spots of forest loss to identify urgent forest loss and synthesize forest-loss data for policy makers, researchers, and journalists who are looking for a concise and comprehensive overview of loss in a particular region. The World Wildlife Fund uses emerging hot spots as an input to identify and report on key deforestation fronts around the world. The Rainforest Trust uses the methodology to identify and prioritize tropical rain forests in dire need of protection through purchases, partnerships, and community engagement interventions.

As more frequent data updates become possible, rapid assessment of the most significant loss can strengthen efforts to reduce deforestation. Rising demand for commodities and a destabilized climate mean that more work is needed to ensure deforestation doesn't continue unchecked, and that we make smart decisions about how we develop and preserve our forests for future generations.

NOTES

Andrienko, G., et al. 2010. "Space, Time and Visual Analytics." *International Journal of Geographical Information Science* 24, no. 10: 1577–1600.

Hansen, M. C., et al. 2013. "High-Resolution Global Maps of 21st-Century Forest Cover Change." *Science* 342: 850–53.

Harris, N. L., et al. 2017. "Using Spatial Statistics to Identify Emerging Hot Spots of Forest Loss." *Environmental Research Letters* 12, no. 2.

Molinario, G. 2015. "Forest Cover Dynamics of Shifting Cultivation in the Democratic Republic of Congo: A Remote Sensing Based Assessment for 2000–2010." *Environmental Research Letters* 10, no. 9.

Ord, J. K., and A. Getis. 1995. "Local Spatial Autocorrelation Statistics: Distributional Issues and an Application." *Geographical Analysis* 27, no. 4: 286–306.

Rew, R., and G. Davis. 1990. "NetCDF: An Interface for Scientific Data Access." *IEEE Computer Graphics and Applications* 10, no. 4: 76–82.

Turubanova, S., et al. 2018. "Ongoing Primary Forest Loss in Brazil, Democratic Republic of the Congo, and Indonesia." *Environmental Research Letters* 13, no. 7: 074028.

Data, maps, and community engagement can lead to positive outcomes for forests and the people who rely on them. (Photo Courtesy of Center for International Forestry Research.)

HOW TO FEED THE WORLD

Agricultural science is searching for more efficient and sustainable farming practices in the face of forecasts that call for an additional 2 billion mouths to feed by 2050. Big data from new innovations in sensors, delivered within the geospatial cloud, will in turn enable a new crop of precision farming techniques and analytics.

By Daniel Roberts, Bruce Vandenberg, Steven Mirsky, and Michael Buser, **USDA—Agricultural Research Service;** Chris Reberg-Horton, **North Carolina State University—Center for Environmental Farming Systems;** and Nick Short and Sudhir Shrestha, **Esri**

Milling truck in a field harvesting crops near Austin, Texas. Access to adequate healthy, nutritious food is central to several UN Sustainable Development Goals.

THE CHALLENGE CONFRONTING AGRICULTURE AND THE WORLD

Regardless of the country or farming region, feeding a human population that could reach 10 billion by 2050 presents enormous challenges to the agricultural community.[1] Plant-food production must increase 60 percent to 100 percent to keep pace with current food consumption, food waste levels, and population trends. Food producers must successfully grow more crops, and while enacting strategies to conserve, they must also build soil resources and minimize agriculture impacts on the environment. Additionally, they must enhance the nutritional quality of many plant foods to ensure a healthy human population.[2,3] New agricultural practices required to meet this challenge must evolve quickly because a radical change in farm practices can take decades to adopt.

Simply increasing land acreage devoted to crop production will not likely satisfy the food needs of future populations. Competition for land use with urban development and the loss of land to salination and desertification will reduce suitable farming land. Repurposing natural landscapes for farming also impacts global carbon and hydrological cycles, greenhouse gas emissions, soil conditions, and biodiversity. It's also unlikely that agriculture can increase crop production using previous agricultural intensification methods. Food production doubled worldwide during the past several decades, largely because of the extensive use of synthetic fertilizer, pesticides, and irrigation.[4,5,6]

These methods are unsustainable for a few reasons. First, certain feed stocks for fertilizer production are dwindling, making their future availability uncertain. Second, water scarcity has affected many areas around the world, and the problem is expected to worsen. In addition, agricultural intensification practices during the past decades have increased soil erosion and decreased soil fertility; polluted groundwater; contributed to eutrophication of rivers, lakes, and coastal ecosystems; and increased greenhouse gases.[4,5,6] These harmful impacts on soil, water, and the atmosphere will dramatically affect food production going forward. Clearly, we must transform agricultural systems to scale up food production while reducing environmental impacts on a finite amount of land.

Food producers also must expand the development of crop cultivars and their use in crop production systems. (A cultivar—cultivated variety—is an assemblage of plants developed for specific beneficial characteristics.) Public and private breeding programs have focused on traits affecting yield, pest and disease resistance, and appearance rather than traits affecting nutritional composition of edible plant parts.[7] Too many farmers have poorly aligned their crop and cultivar choices with human dietary needs, understandably being driven mostly by price, yield, and market preference.[8] As a result, crop cultivars tend to be calorie rich—containing macronutrients such as fat, protein, and carbohydrates—but poor in vitamin and mineral micronutrients and in human health-promoting phytochemicals. A key technical challenge for food producers is thus to expand crop cultivar development and use to maximize yield and nutritional quality. Crop cultivars must also be developed to efficiently use water and soil nutrients needed for plant growth.[9,10]

Global climate change will greatly constrain our ability to produce more nutrient-dense food.[1] Researchers expect climate change to bring increased temperatures and increased frequency and intensity of extreme weather and drought. These changes may well offset any possible benefits to crop yield because of expected associated negative impacts. Yields of most crops decline dramatically at temperatures much above 30 degrees Celsius. Drought, salinity stress, higher ozone levels, and the onset of new pest and pathogen problems would limit crop yields.[11] And elevated temperature and carbon dioxide levels can degrade the nutritional quality of certain crops.[12]

Getting the most nutrition from the world's agricultural operations is inherently a geospatial problem: crops grow in specific places with specific climate and soil conditions. Within individual fields, advanced data collection methods now allow farmers to alter watering and fertilization practices at a precise level within a single crop. Thanks to advances in geographic information systems (GIS) and real-time machine learning, farmers can now analyze this explosion of new crop data to maximize the efficiency of their operations, even to the point of making real-time adjustments.

Fertilizer and eroded topsoil spill from an unprotected Iowa farm in a rainstorm.

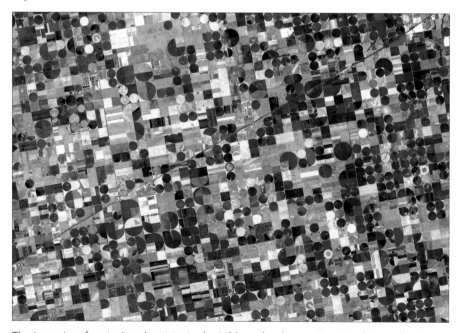

The intensity of agricultural activity in the US heartland is apparent in this satellite image of circular crop fields in Haskell County, Kansas.

GIS: HELPING CHANGE THE WAY WE GROW FOOD

Managing data spatially is inherent to the development of next-generation crop production systems.[13] Agricultural fields are highly heterogeneous with respect to properties that impact plant growth and health. Topography, soil type, and pathogen and pest populations can vary considerably within and among farmers' fields. These field properties interact with climate, greatly influencing drainage, water and nutrient availability, and pest outbreaks and their spatial distribution. Using GIS, we can spatially organize geospatial data from sensors for crop yield, soil fertility factors, water, pathogens, and so on. Food producers can link the spatial patterns of field properties with climate factors to develop correlations between crop yield, soil type, and crop and soil management. Further, they can use this information to more precisely manage varying crop populations and application rates of fertilizers, pesticides, and irrigation. As a result, farmers can stop treating crop production fields as a uniform management unit and instead treat the field as a heterogeneous substrate for plant growth. In this way, they can maximize crop productivity and profitability and more efficiently use inputs of resources (fertilizers, pesticides, water), resulting in less loss and fewer harmful environmental impacts. Scientists developed this precision agricultural approach in the 1980s with the advent of global positioning system (GPS) technology and GIS and improved it with sensor technology, big data approaches, algorithms, and robotics.[13] With increased farmer adoption of precision agriculture (e.g., adoption was only 30–50 percent on US-grown corn and soybeans in 2012),[14] we can minimize the footprint of food production on the environment going forward.

Environmentally benign crop production tools (i.e., cover crops and beneficial microbes) are being developed to use in next-generation crop production systems to make food production more sustainable.[3] Farmers can use these biologically based tools in place of, or in conjunction with, reduced amounts of synthetic fertilizer or pesticides. Cover crops provide a diverse array of benefits and are gaining traction among US growers, their use being highly incentivized by state, federal, and private programs. Farmers grow cover crops during fallow periods in a crop rotation. They typically plant them in the fall after the summer cash crop and terminate them in the spring before planting the next cash crop. Cover crops fix atmospheric carbon dioxide and build soil organic matter as they decompose and contribute to soil health. Cover crops also capture excess nutrients and prevent nutrient loss from soil to aquifers and waterways, prevent soil erosion, and increase rainfall infiltration and soil water-holding capacity. In some cases, cover crops help control weeds, pathogens, and insect pests. Leguminous cover crops such as hairy vetch and clover fix atmospheric nitrogen and thereby improve soil fertility.[15] Farmers can use other biologically based crop production tools, such as beneficial microbes, in next-generation crop production systems to control plant disease and insect pests, enhance soil phosphorus availability, and promote plant growth.[3]

To optimize the benefits of biologically based crop production tools, farmers must integrate them into a precision framework to facilitate site-specific decision-making, an approach called *precision-sustainable agriculture*. Use of these biologically based tools adds another layer of management complexity. Their use integrates many factors driving the performance of cover crops and microbes and impacting the performance of cash crops. Using these tools in a precision framework is a challenge because data must be in an actionable state for real-time decision-making. The ability to manage large amounts of data in precision-sustainable agriculture provides farmers with timely, specific, and context-appropriate information and is key for grower adoption. Geospatial information tools offer this ability because they manage complexity at scale.[13]

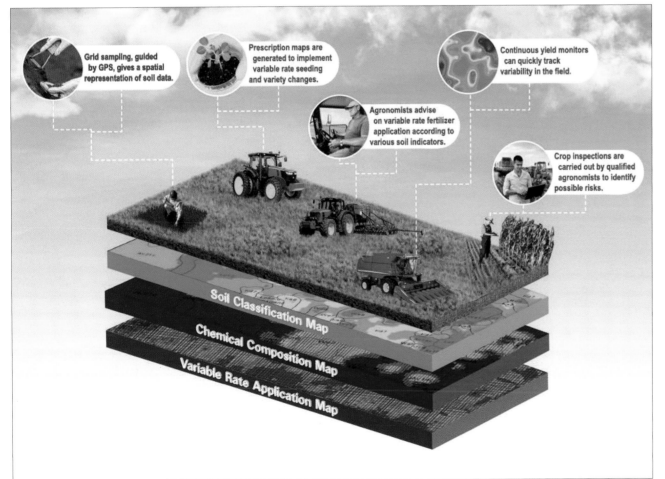

Grid sampling, guided by GPS, gives a spatial representation of soil data.

Prescription maps are generated to implement variable rate seeding and variety changes.

Continuous yield monitors can quickly track variability in the field.

Agronomists advise on variable rate fertilizer application according to various soil indicators.

Crop inspections are carried out by qualified agronomists to identify possible risks.

Soil Classification Map

Chemical Composition Map

Variable Rate Application Map

Precision agriculture production systems rely heavily on the fusion of volumes of data from a variety of data generators with varying degrees of data velocity and moving between many user nodes.

CASE STUDY: COVER CROPS AND THE SMART FARM

Work with cover crops illustrates how GIS underpins technology created to optimize performance and speed adoption in the agricultural community. Crucial advances in data science, sensing technology, and artificial intelligence (AI) and machine learning (ML) applications have ignited a digital revolution, which in turn has led to adaptive, geospatial decision support tools and on-farm monitoring systems. These systems provide real-time data for the most effective use of cover crops, which are grown primarily to protect and improve the soil, as opposed to cash crops, which are grown for their commercial value.

Cover crops respond to environmental conditions in fields that impact crop biomass and quality—two performance factors tightly correlated with benefits that cover crops deliver to these fields.[16,17] New sensing technologies coupled with AI algorithms can quantify the performance factors and spatial variability of cover crops and how they affect cover-crop benefits. Remote sensing offers many tools relevant to cover crops, including populating landscape-level models with estimates of percentage of land cover. Scientists have used the normalized difference vegetation index (NDVI) based on reflectance of cover crops to evaluate soil cover and cover-crop biomass.[18,19] Farmers are increasingly using cover-crop mixes to deliver multiple benefits, which requires cover-crop species identification for optimal performance. Scientists also frequently use lidar technology to map structure, including vegetation height, density, and other characteristics across a region. Lidar helps scientists quantify plant height and biomass for cover crops and weeds. Lidar data and red/green/blue (RGB) digital images also help farmers identify species.

New geospatial decision-support tools using models calibrated from imagery for farm-specific conditions integrate data from these on-site sensors. These tools take real-time environmental conditions and other site-specific factors into account in offering recommendations to farmers.[20] Many current decision-support tools suffer from usability issues that stymie adoption of the tools and practices they support. Going forward, decision-support tools must be designed to reduce information overload on users, use realistic modeling techniques that integrate field data, and enable site-specific decision-making.[21] Providing information and tools for processing this information in near real time will allow farmers to optimally adjust the way they manage cover and cash crops, thus increasing profits and reducing stress.

Going forward, tools must be designed to move on-station field experiments to the farm for development of decision-support tools. Historically, approaches to manipulate farmer practices in their fields have been met with varied success, often failing 50 percent of the time. This success rate speaks mostly to the complexity of farming and how farmers must react to climate and logistics in real time. Fortunately, a diverse array of sensing platforms makes it possible to link large networks of farms to exploit the communal nature of farmers and to explore causal relationships between climate, soil, and farm management without additional cognitive time burdens to farmers. Increasingly, farmers—even small family farmers—are becoming comfortable with technology.

These next sections will show how to use geospatial tools to manage and monitor the performance of cover crops and cropping systems in general.

Today's modern farmer is at home with technology. The ubiquity of digital communications means everyone already knows how to use these tools.

Inexpensive tablet technology brings the "agri-data" to the field.

The Beltsville research farms complex

Located less than 20 miles from the nation's capital, the Beltsville Agricultural Research Center (BARC) is actually a collection of growing fields and research facilities located around Beltsville, Maryland. Scientists there lead the operation of working farms to envision, create, and improve agricultural knowledge and technologies. The center's mission is to help the United States and the world provide healthy crops and animals; clean and renewable natural resources; sustainable agricultural systems; and abundant, high-quality, and safe agricultural commodities and products.

The Beltsville complex is managed by the US Department of Agriculture's (USDA's) Agricultural Research Service (ARS). Overall, ARS is the chief scientific in-house agricultural research agency for the nation. ARS scientists perform research on more than 660 projects within 15 national programs. Their research covers crops, insects, animals, nutrition, fertilizers, water, and many other broad topics that help create a viable food supply for a nation of more than 300 million people.

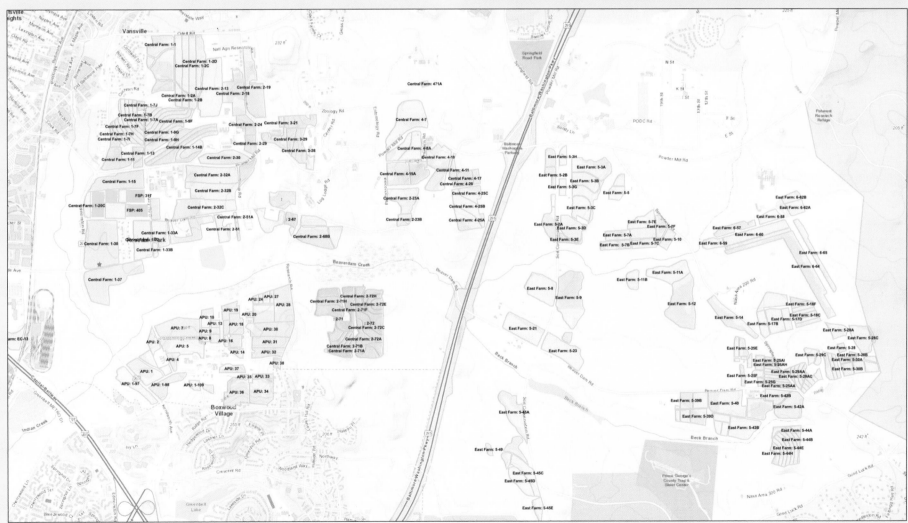

ARS scientists perform much of their work in a complex of farm plots named for Henry A. Wallace, America's 11th Secretary of Agriculture and former US vice president.

Application of Earth observation data for cover-crop spatial variability

Collecting ground truth is key not only for traditional models but also for developing training data for AI/ML, which is heavily dependent on collecting large amounts of training data for the models to properly generalize. Reducing training data collection costs through a variety of techniques is, therefore, critical for driving the adoption of these modern models. To illustrate this concept, this section describes the use of GIS, satellite data, ground truth around a cover-crop case study at Beltsville Argricultural Research Center (BARC) using a variety of technologies designed to reduce data collection times in the context of the desktop GIS software ArcGIS® Pro. The case study used VENµS (vegetation and environmental monitoring on a new microsatellite) imagery for the exploratory analysis in ArcGIS Pro that feeds into a data network used by cover-crop researchers.

The case study used 13 cloud-free images for BARC available from November 2018 to May 2019. These superspectral images have 12 visible near-infrared spectral bands, a swath width of 27.6 kilometers, and spatial resolution of 10 meters. A mosaic dataset—a data model developed by Esri within the geodatabase to manage a collection of raster datasets (images)—allows one to store, manage, view, and query large collections of raster and image data. At the start of the process, an empty mosaic dataset was created as a container in the geodatabase in ArcGIS Pro that was populated with the VENµS data.

Cover crops in the test field at Beltsville respond to environmental conditions that impact biomass and quality. The NDVI and Soil Adjusted Vegetation Index (SAVI) were used as good estimators in aboveground biomass. Raster functions that allow on-the-fly processing operations were used to explore spatial variation of vegetation health and to understand the relationship between soil moisture and vegetation. These indices were saved as persisted cloud raster format (CRF) where the CRF data cube is optimized for writing and reading large files in a distributed processing and storage environment. In a CRF file, large rasters are broken down into smaller bundles of tiles, allowing multiple processes to write simultaneously to a single raster.

The VENµS microsatellite

The Israeli VENµS program is a vegetation and environment monitoring system that utilizes emerging microsatellite technology. The sensor provides high-resolution digital multi- and superspectral imagery to monitor, analyze, and model land surface behavior under different parameters. This is high spatial resolution Earth imaging for a wide range of commercial and scientific applications.

Artist's rendering of a VENµS microsatellite in action, with the rainbow suggesting the multiple spectral bands "visible" to the sensor.

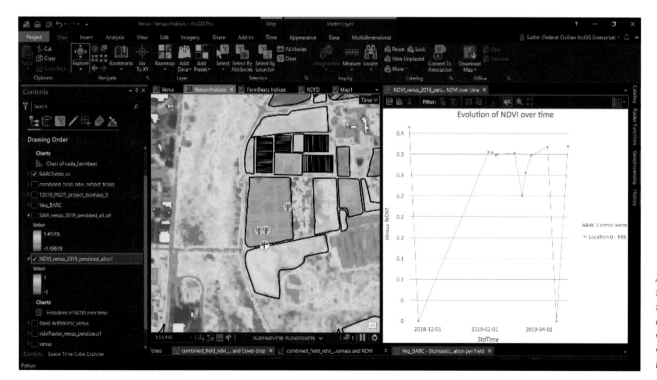

ArcGIS Pro map and chart of NDVI readings from the VENµS microsatellite. NDVI is a standard way to measure healthy vegetation. High NDVI values correspond to healthier vegetation. Low NDVI values correspond to less or no vegetation. This view of the BARC test field data shows changes in NDVI during the winter cover-crop period.

Ground truth

In remote sensing, *ground truth* refers to information collected on location, which can be compared to remote sensing data (e.g., data collected from VENµS) for validation. The acquisition of carefully documented ground-truth data enables scientists to calibrate their models and aids in the interpretation and analysis of what is being sensed. To illustrate the correlation between actual biomass and the VENµS vegetation indices for biomas estimation, researchers intially collected ground truth manually at randomly dispersed locations (shown as red dots on the Central Farm 4-7 map). This literal "on the ground" data was then supplemented with image data collected from a multispectral, camera-bearing Hiboy tractor system, which collected NDVI samples directly.

Driving a Hiboy tractor equipped with sensors through a soybean field at the USDA research farm in Beltsville, Maryland.

However, since not every farmer has a Hiboy platform with an NDVI sensor at their disposal, an emerging investigative area at BARC is the deployment of low-cost robotics. These so-called Internet of Things (IoT) approaches offer lower-cost options moving forward that might ultimately lead to the Holy Grail for farming: an automated recommendation engine that growers can act on in real time. The EarthSense robot, called TerraSentia, has been developed to measure attributes such as plant height, stand counts, stem widths, and so on for under-canopy plant phenotyping. By navigating through a cornfield, for example, the robot can use methods ranging from computer vision to 2D lidar to construct a model that can be associated with field maps to support biomass volume estimation. Therefore, a swarm of TerraSentia robots operating simultaneously in a field offers an efficient method for continuously monitoring the growth of biomass.

TerraSentia is an automated, less-expensive robotic solution that uses computer vision and lidar to get detailed biomass estimates for use as ground truth.

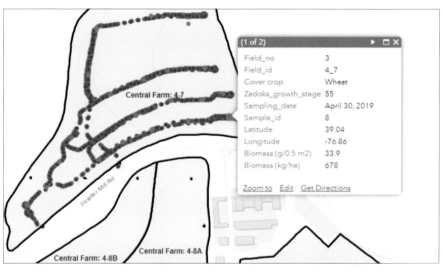

Central Farm 4-7 map showing the sample point locations and a pop-up indicating biomass and other details from a data collection pass on April 30, 2019.

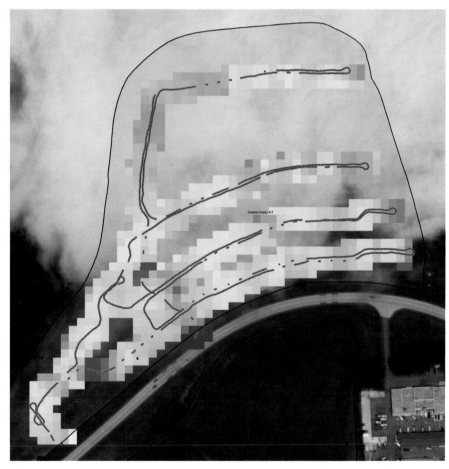

This map shows the result of manually collecting groundtruth biomass, which can be used for calibration. The map also shows the path of the Hiboy through the same field to collect many more control points. By averaging the Hiboy points using a nearest-neighbor approach, we can compare the results directly to the VENµS reflectance values.

Crop field boundary detection using deep learning

The previous section focused on cost-effective methods for collecting ground truth and correlating this with well-known vegetation indices using out-of-the-box functionality in ArcGIS Pro. But new capabilities in ArcGIS Pro using ML and AI will significantly reduce the time it takes to build these models.

To develop these new capabilities, ARS researchers considered one of their classic problems: how to define and create features depicting field boundaries (i.e., crop masking), called common land units. Using automated techniques has the potential to significantly reduce the time needed for farmers to create reports on their current fields and crop plantings, as typical techniques require looking over multiple growing seasons for field accuracy. But this work also shows great promise in helping scientists build potential candidate farms for their on-farm research network, and acting as geofences for queries to imagery stores and other data sources.

Farmers use the same techniques to identify crop types that are used to quantify global production of cash crops such as corn, soybeans, and so forth, and this is essential information to people working in the commodities markets.

Using the Beltsville farm as an example, scientists classified VENµS imagery to determine field boundaries. Specifically, the imagery was resampled to 1 meter to match the spatial resolution of the National Agriculture Imagery Program (NAIP) output. The images are band stacked to get 16-band imagery (12 bands of VENµS and 4 bands of NAIP) using the Composite Bands geoprocessing tool as shown in the NAIP imagery. The trained neural network model has two outputs (croplands versus noncroplands). The NAIP imagery shows the workflow in ArcGIS Pro.

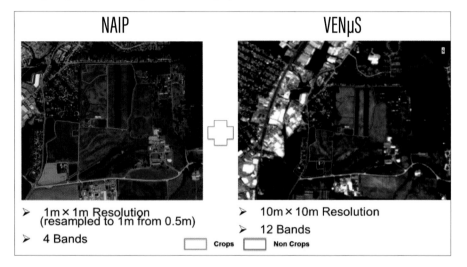

NAIP
- 1m × 1m Resolution (resampled to 1m from 0.5m)
- 4 Bands

VENµS
- 10m × 10m Resolution
- 12 Bands

Crops | Non Crops

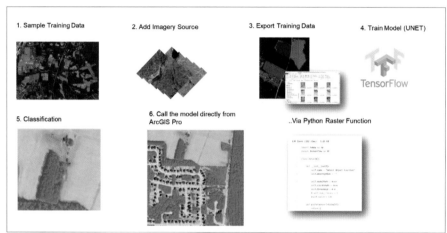

Aerial platforms capture NAIP imagery from USDA at high resolution and cover the entire United States every two years. Software training data for classification of the VENµS data is created with ArcGIS Pro.

Class	Precision %
Crops	90.2
Forest	92.6
Impervious	85.4
Water	88.3

ArcGIS Pro and machine learning can classify VENµS data using NAIP imagery. The resultant image clearly delineates crop fields in black versus other features such as forest and water.

Field data to analytics: Real-time data flow for cover crops

Microsoft Azure's FarmBeats is another way to collect and aggregate ground-truth data. The product combines IoT sensors, data analysis, and ML on a cloud-based framework and allows farmers to gather data-driven insights in a cost-effective way. FarmBeats uses AI/ML to turn data from many sources, including sensors, satellites, drones, and weather stations, into actionable intelligence for farmers. FarmBeats also uses a new, low-cost networking solution to connect the sensor network to the cloud. The connection uses unused frequencies allocated to broadcasting services (called TV white space) not used in rural areas.

Combining FarmBeats with the ArcGIS platform creates the potential to develop a complex network of sensors from the agriculture industry with GIS data. For the cover-crop network, a Python script pulls data from the FarmBeats application program interface (API) for each sensor location and populates the data to ArcGIS Online as a feature service every 15 minutes.

This real-time data pipeline opens the door for field data analytics and visualization, improving the access to spatial analysis tools in the ArcGIS platform. As an example of a decision-support tool, the team built a dashboard on top of real-time data to monitor a kind of digital "heartbeat" of field observations, a critical tool for the modern farmer.

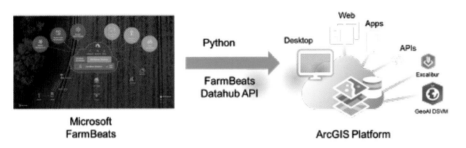

Visualizing the FarmBeats data with the ArcGIS platform.

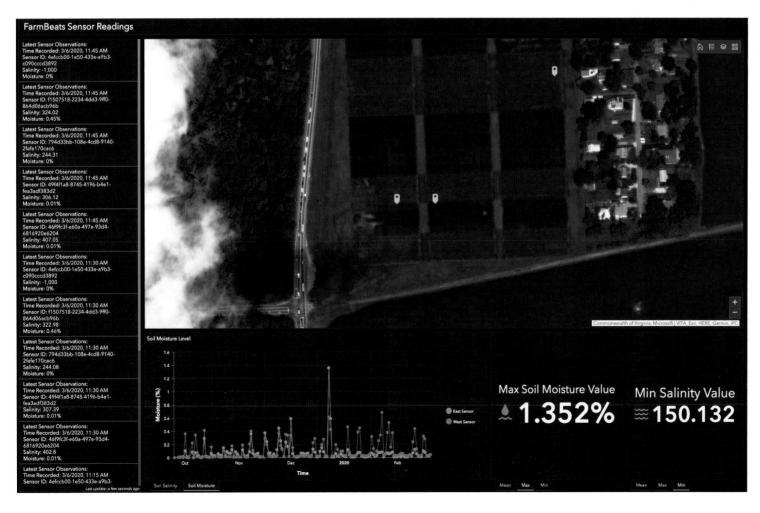

GIS-driven dashboards allow data from FarmBeats and other sensors to display in real time.

AUTOMATING AND ASSISTING THE FIELDWORK OF CROP DEVELOPMENT

GIS serves as the backbone for emerging technologies that support the development of cultivars with increased resilience to abiotic and biotic stressors related to climate change and nutrition. For example, drought—the greatest threat to soybean yields—is expected to more than double worldwide by 2050. For this reason, developing a more drought-resistant soybean cultivar is a priority. The genetics of drought tolerance are complex and poorly understood, so field-based phenotyping is typically central to soybean cultivar development. Although visual traits that correlate with drought response are known for soybeans (e.g., midday leaf wilting), screening for these traits is time consuming and often requires highly trained screeners. This reality limits the scale and number of testing locations.

A system that automatically rates drought-related phenotypes in soybeans in a high through-put scale would significantly reduce the time and labor required to increase the accuracy of soybean cultivar selection. Coupling AI with machine vision, researchers can use low-cost solutions such as simple RBG imaging taken with drone-mounted cameras to detect drought stress. This method can be widely distributed across breeding programs to remove the subjective nature of human scoring systems and greatly accelerate producing drought-resistant soybean lines. When the camera network detects stress, it can tansfer images of the affected plots to scientists. Categorizing stress levels within the camera requires relatively small data streams across rural cellular connections.

FarmWave® is a good example of this technology because it focuses on identifying and diagnosing pests and pathogens that impact crops. FarmWave uses cell phone technology to capture geocoded ground-truth data that can be shared on a common platform with the producer and scientific community.

Speech recognition technology also plays a pivotal role in crop development in the form of AgVoice®, a voice recognition technology. With AgVoice, the application uses an AI *bot* to prompt you through data collection. The focus is a two-way interaction on domain-specific knowledge around food and agricultural terms.

GIS: Helping change the way we do science

The rapid development and transfer of these new technologies will help farmers produce more food with better nutrition. Yet these technologies must safeguard the integrity of the soil, water, and air needed to grow this food while also confronting the headwinds of global climate change. Until recently, the public and private sectors largely developed technologies in silos, slowing progress. However, recent advances in computing infrastructure, big data, and advanced algorithms portend a paradigm shift in the way we develop technology in the agriculture sector. An infrastructure based on these new technologies will allow the mass transfer and sharing among scientists of agronomic data required for the development of next-generation cropping systems and omics (genomic, transcriptomic, proteomic, metabolomic) and other data necessary to develop advanced crop cultivars. Integrating data collected from sensored smart farms into a collaborative network will create a positive feedback loop that allows rapid testing of next-generation cropping systems and advanced crop cultivars for different crop-production regions worldwide. On-station laboratories now must move to farms to better represent farm conditions and subtle differences in farming practices. A diverse array of sensing platforms can link large networks of farms to explore causal relationships between climate, soil, and management. The ability to analyze data from more farming sites results in better models, decision-support tools, and outcomes, and will result in more growers adopting these methods.

Like virtually all human activities, corn farming is being massively disrupted by the digital revolution.

Technologically, providing a platform for connecting field systems and existing GIS nodes results in a new architectural pattern called Web GIS. The Web GIS pattern supports implementation patterns ranging from on-farm, edge-oriented architecture to the *system of systems* approach that acts like a nervous system for agriculture. From an agricultural perspective, Web GIS provides a framework for reducing silos between scientists, across the public sector, and between the public sector and the agricultural industry.

Farmwave app running on Apple iOS.

AgCROS: A collaborative data-sharing platform

The USDA uses GIS built on a cloud infrastructure to increase collaboration among ARS scientists and collaborators to enhance cooperative science. Developed using the industry standard *Agile* approach, the Agricultural Collaborative Research Outcome System (AgCROS) illustrates a collaborative vision by providing a single platform to store and disseminate new agricultural data and models.[22] Data from studies on greenhouse gas emissions, soil health, genomics, cover crops, renewable energy, antibiotic resistance, nutrient use, and nutrition is all contained within AgCROS.[22,23] AgCROS is built on ArcGIS Hub with connections to ArcGIS Online and ArcGIS Enterprise. Data for the system is stored on Microsoft Azure ARS cloud as an enterprise geodatabase and referenced in ArcGIS Online. ArcGIS Enterprise allows for storage of imagery, and real-time data. The ARS cloud allows for on-demand processing of AI and ML techniques and other analytical tools. Microsoft FarmBeats and Esri Geo-event server combine to act as the sensor data generator. Both systems use IoT to handle the variety of sensor types that will be deployed as well as data types. TV white space antennas serve as the way of transmitting sensor data to the cloud in areas without internet connectivity.

Remote-sensed imagery data from satellites, planes (lidar), and land or air drones requires large amounts of storage that the cloud allows for via hard-drive scalability. In addition to the automated ground-truth data-collection techniques discussed in the previous sections, Esri's field mobility apps such as ArcGIS Survey123 and ArcGIS Collector are standards built into AgCROS to systematically reduce collection times and increase data quality for field scientists and collaborators. These applications gather data on mobile devices whether internet connections exist or not. Data gathered on these applications mobile-sync with AgCROS when internet connectivity is available. The applications provide ground-truth GIS data, and users can customize them to take pictures, scan barcodes, and gather a plethora of other observation data that does not have sensor inputs.

Having all the data and tools in the GIS central nervous system allows users to compare and extrapolate AI/ML worldwide at speeds previously not possible. These advances allow users to consider ways to improve farming methods while reducing harmful environmental impacts.

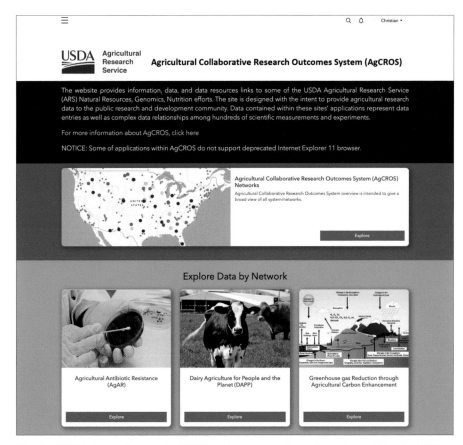

The ArcGIS Hub interface to AgCROS, which covers the various sites within the ARS research network.

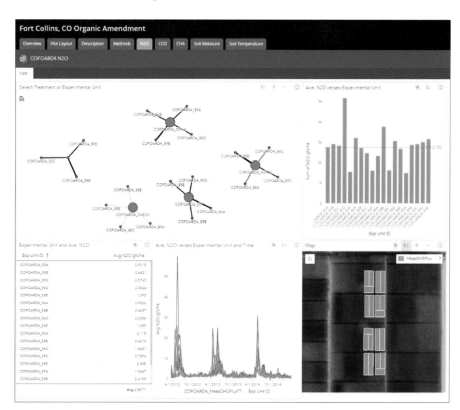

Scientists can drill down from a site to the field level as shown. Once at the site level, scientists can use various tools such as graphing across sites to get an overall view of the data in question, in this case nitrous oxide.

Breaking silos with services-oriented architecture

The big challenge concerning Earth observations data in an agriculture domain is how to make complicated data available and interoperable for a growing user community with varying interests and domain expertise. Based on the Web GIS pattern, the ArcGIS platform is evolving in step with the larger technology industry to help meet these challenges.

In recent years, changes in the geospatial community have required more collaboration. The growth of cloud computing also has supported the use of scientific data from real-time field observations with sensors, field data collection, and drone observation. Most GIS analysts and remote sensing scientists once worked on their own projects, data, and computers. But the growing volume and diversity of scientific data and evolution of technology illustrated the benefits of sharing data and methods. Scientists developed centralized data storage and some centralized analytic services for use within their operations. Many organizations adopted a *services-oriented architecture* (SOA).

Collaborative communities have emerged that reach beyond corporate or organizational boundaries.[24] This system of systems is simply a collection of portals containing distributed data and distributed analytics, which in turn can interoperate as a single system. This breakthrough allows people to collaborate across space and discipline. One example of this is the Global Earth Observations System of Systems

(GEOSS), which has 105 member countries and many affiliated research groups, including Esri and many of its customers active in GEOSS. Similarly, as a network of networks, the AgCROS bridges earlier gaps in data and analytics.

Sharing the data and models through web services increases the repeatability of consistent and reproducible research workflows. These developments, now in practical use throughout USDA, have greatly improved collaboration. Cloud computing in agricultural science has enabled scientists to put their focus where it matters most: improving precision agriculture.

With the growth of cloud computing, a platform such as ArcGIS allows scientists to deliver their unique data and analytics to any desktop and push it back to the web. Web GIS brings analytics to spatial data in a new way. Researchers previously had to process, modify, and extract data to answer a set of questions. Web GIS transforms data into web maps and services that are mashed up with different layers, so that data can answer questions dynamically without processing them for each parameter. Web GIS in the hands of a much larger audience reduces the need to create custom applications, provides a platform for integrating GIS with other business systems, and enables cross-organizational collaboration.

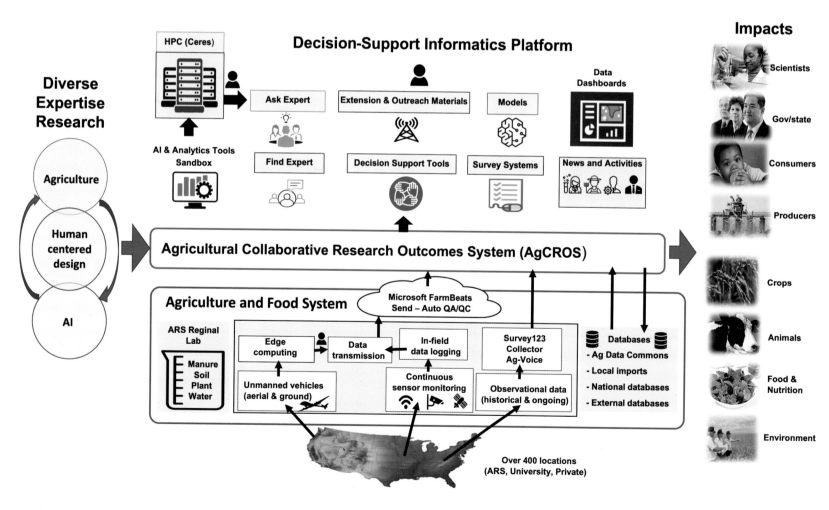

CONCLUSIONS

GIS plays a major role in development of next-generation cropping systems and crop cultivars that are more nutritious and more resilient to biotic stress, abiotic stress, and other factors associated with climate change. Perhaps most important, scientists now can use GIS to enhance collaboration and speed development of new technologies. Future plans include linking ARS databases and modeling environments in AgCROS with databases and other resources from sister USDA agencies so that farmers and other members of the agricultural community can more readily access and use their services.

NOTES

1. H. C. J. Godfray et al., "Food Security: The Challenge of Feeding 9 Billion People," *Science* 237 (2010): 812–18.

2. J. A. Foley et al., "Solutions for a Cultivated Planet," *Nature* 478 (2011): 337–42.

3. D. P. Roberts, and A. K. Mattoo, "Sustainable Agriculture—Enhancing Environmental Benefits, Food Nutritional Quality and Building Crop Resilience to Abiotic and Biotic Stresses," *Agriculture* 8 (2018): 8. doi.10.3390/agriculture9010008.

4. J. A. Foley et al., "Global Consequences of Land Use," *Science* 309 (2005): 570–74.

5. P. A. Matson et al., "Agricultural Intensification and Ecosystem Properties," *Science* 277 (1997): 504–09.

6. D. Tillman, et al., "Forecasting Agriculturally Driven Global Environmental Change," *Science* 292 (2001): 281–84.

7. D. C. Sands et al., "Elevating Optimal Human Nutrition to a Central Goal of Plant Breeding and Production of Plant-Based Foods," *Plant Science* 117 (2009): 377–89.

8. R. A. Halimi et al., "Bridging the Food Security Gap: An Information-Led Approach to Connect Dietary Nutrition, Food Composition, and Crop Production," *Journal of the Science of Food and Agriculture* (2019), doi.10.1002/jsfa.10157.

9. A. K. Mattoo, "Translational Research in Agricultural Biotechnology—Enhancing Crop Resistivity against Environmental Stress alongside Nutritional Quality," *Frontiers in Chemistry* 2 (2014): 30, doi: 10.3389/fchem.2014.00030.

10. D. P. Roberts and A. K. Mattoo, "Sustainable Crop Production Systems and Human Nutrition," *Frontiers in Sustainable Food Systems* 3 (2019): 72, doi. org/10.3389/fsufs.2019.00072.

11. N. V. Federoff et al., "Radically Rethinking Agriculture for the 21st Century," *Science* 327 (2010): 833–34.

12. S. S. Myers et al., "Climate Change and Global Food Systems: Potential Impacts on Food Security and Undernutrition," *Annual Review of Public Health* 38 (2017): 259–77.

13. J. A. Delgado et al., "Big Data for Sustainable Agriculture on a Geospatial Cloud Framework," *Frontiers in Sustainable Food Systems* 3 (2019): 54, doi: 10.3389/fsufs.2019.00054.

14. D. Schimmelphfenning, "Farm Profits and Adoption of Precision Agriculture," 2016, https://www.ers.usda.gov/webdocs/publications/80326/err- 217.pdf?v=0, USDA ERS.

15. A. K. Mattoo and J. R. Teasdale, "Ecological and Genetic Systems Underlying Sustainable Horticulture," *Horticultural Reviews* 37 (2010): 331–62.

16. D. M. Finney, C. M. White, and J. P. Kaye, "Biomass Production and Carbon/ Nitrogen Ratio Influence Ecosystem Services from Cover Crop Mixtures," *Agronomy Journal* 108 (2016): 39–52.

17. S. B. Mirsky et al., "Characterizing Cereal Rye Biomass and Allometric Relationships across a Range of Fall Available Nitrogen Rates in the Eastern United States," *Agronomy Journal* 109 (2017): 1520–31.

18. W. D. Hively et al., "Using Satellite Remote Sensing to Estimate Winter Cover Crop Nutrient Uptake Efficiency," *Journal of Soil and Water Conservation* 64 (2009): 303–13.

19. K. Prabhakara, W. D. Hively, and G. W. McCarty, "Evaluating the Relationship between Biomass, Percent Groundcover, and Remote Sensing Indices across Six Winter Cover Crop Fields in Maryland, United States," *International Journal of Applied Earth Observation and Geoinformation* 39 (2015): 88–102.

20. M. J. O'Grady and G. M. P. O'Hare, "Modelling the Smart Farm," *Information Processing in Agriculture* 4 (2017): 179–87.

21. S. E. Cook and R. G. V. Bramley, "Is Agronomy Being Left Behind by Precision Agriculture?" in *Proceedings of the 10th Australian Agronomy Conference* (Hobart, Tasmania: The Australian Society of Agronomy, 2001).

22. J. A. Delgado et al., "Agricultural Collaborative Research Outcomes System: AgCROS–An Emerging Network of Networks for National Food and Environmental Security and Human Health," *Journal of Soil and Water Conservation* 73 (2018): 158A–64A.

23. S. J. Del Grosso et al., "Introducing the GRACEnet/ REAP Data Contribution, Discovery, and Retrieval System," *Journal of Environmental Quality* 42 (2013): 1274–80.

24. S. R. Shrestha et al., "Earth Observation and Geospatial Implementation: Fueling Innovation in a Changing World," in *Earth Observation Open Science and Innovation*, ed. P. P. Mathieu and C. Aubrecht, ISSI Scientific Report Series, vol. 15. (Cham: Springer, 2018).

PART 3
HOW WE LOOK AT EARTH

Successfully understanding how Earth works and how Earth looks to us requires integrative and innovative approaches to observation and measurement. These approaches include Earth observation in varying forms, such as from sensors on satellites, aircraft, drones, ships, and so on. They also include the important data science issues of conducting analysis; modeling, developing, and documenting useful datasets for science; and interoperating between these datasets and between various approaches.

MONITORING AIR QUALITY IN THE UNITED STATES

Born in the midst of rising concern about harmful pollution in 1970, the US Environmental Protection Agency (EPA) has since its inception focused considerable attention and resources on air quality. Using GIS, the agency's Office of Air Quality Planning and Standards compiles, synthesizes, and publishes data to guide policy that keeps the public safe and informed.

By Liz Naess and Halil Cakir, **EPA;** and Alberto Nieto, **Esri**

In 1973, Los Angeles, California, had some of the worst air pollution in the nation. Efforts by the EPA to document the problem led to greater awareness, then to legislation, and ultimately to regulations such as the catalytic converter that noticeably reduced particulate air pollution.

THE VALUE OF AIR QUALITY

When was the last time you considered the quality of the air that you breathe?

Many of us take for granted the quality of our air that allows us to live our daily lives in relative comfort. Most of us can go out for a morning walk and enjoy outdoor activities without constantly checking air quality monitors for hazardous levels of pollutants.

Is poor air quality a thing of the past? Some of us may have heard of or even experienced the Donora Fog. On October 27, 1948, a sudden onset of smog settled over the town of Donora, Pennsylvania. Sulfur dioxide emissions from US Steel's Donora Zinc Works and its American Steel & Wire plant occurred frequently in Donora. But on that day in 1948, a temperature inversion trapped a mass of warm, stagnant air in the valley. Pollutants in the air mixed to form a thick, yellowish, acrid smog that blanketed Donora for five days. When it finally cleared, 20 people had died, and the smog sickened more than 6,000 people.

Donora was not the only place experiencing air pollution with deadly consequences: Almost all major US cities routinely experienced toxic air. Even though they were not recognized by health officials immediately, smog events killed hundreds of people in 1953 and 1966 in New York. Elsewhere in the world, the "killer fog" of 1952 in London, England, killed an estimated 12,000 people. These events triggered a collective wake-up call that ultimately led to the formation of the EPA in 1970, raised public awareness to the dangers of air pollution, and served as the basis for the Clean Air Act.

Heavy smog shrouds the George Washington Bridge connecting New Jersey and New York. This view faces the New Jersey side of the Hudson River, 1973.

Donora, Pennsylvania, 1948.

A constable at work during London's 1952 "killer fog."

UNDERSTANDING AIR POLLUTION

The Clean Air Act requires the EPA to set National Ambient Air Quality Standards (NAAQS) for six common air pollutants (also known as *criteria air pollutants*) that are common in outdoor air, considered harmful to public health and the environment, and come from numerous and diverse sources. Criteria pollutants are ground-level ozone (O_3), particulate matter (PM), carbon monoxide (CO), lead (Pb), sulfur dioxide (SO_2), and nitrogen dioxide (NO_2).

Ground-level ozone

Health effects: Ozone exposure reduces lung function and causes respiratory symptoms, such as coughing and shortness of breath. Ozone exposure also aggravates asthma and lung diseases such as emphysema, leading to increased medication use, hospital admissions, and emergency department visits. Exposure to ozone may also increase the risk of premature mortality caused by respiratory issues. Short-term exposure to ozone is also associated with increased total nonaccidental mortality, which includes deaths caused by respiratory causes. **Environmental effects:** Ozone damages vegetation by injuring leaves, reducing photosynthesis, impairing reproduction, and decreasing crop yields.

Particulate matter

Health effects: Exposures to PM, particularly fine particles referred to as PM2.5, can cause harmful effects on the cardiovascular system, including heart attacks and strokes. These effects can result in emergency department visits, hospitalizations, and, in some cases, premature death. PM exposures are also linked to harmful respiratory effects, including asthma attacks. **Environmental effects:** Fine particles (PM2.5) are the main cause of reduced visibility (haze) in parts of the United States, including many national parks and wilderness areas.

Carbon monoxide

Health effects: Breathing elevated levels of CO reduces the amount of oxygen reaching the body's organs and tissues. For those with heart disease, this outcome can result in chest pain and other symptoms, leading to hospital admissions and emergency department visits. **Environmental effects:** Emissions of CO contribute to the formation of carbon dioxide (CO_2) and ozone, greenhouse gases that warm the atmosphere.

Lead

Health effects: Depending on the level of exposure, lead may harm the developing nervous system of children, resulting in lower IQs, learning deficits, and behavioral problems. Longer-term exposure to higher levels of lead may contribute to cardiovascular effects, such as high blood pressure and heart disease in adults. **Environmental effects:** Elevated amounts of lead accumulated in soils and freshwater bodies can result in decreased growth and reproductive rates in plants and animals.

Sulfur dioxide

Health effects: Short-term exposures to SO_2 are linked with respiratory effects, including difficulty breathing and increased asthma symptoms. These effects are particularly problematic for asthmatics while breathing deeply, such as when exercising or playing. Short-term exposures to SO_2 have also been connected to increased emergency department visits and hospital admissions for respiratory illnesses, particularly for at-risk populations including children, older adults, and people with asthma. SO_2 contributes to particle formation with associated health effects.

Nitrogen dioxide

Health effects: Short-term exposures to NO_2 can aggravate respiratory diseases, particularly asthma, leading to respiratory symptoms, hospital admissions, and emergency department visits. Long-term exposures to NO_2 may contribute to asthma development and potentially increase susceptibility to respiratory infections.

SETTING AIR QUALITY STANDARDS

The process of reviewing the NAAQS involves assessing new scientific data, understanding the human and welfare impacts from air pollutants, and determining protective levels. Throughout the process, the public has a chance to weigh in on the decisions through public hearings and comment periods.

The Integrated Science Assessment

The Integrated Science Assessment (ISA) is a comprehensive review of policy-relevant science, including key scientific evaluations and causal judgments, which provides the scientific foundation to review criteria pollutants. Draft versions of the ISA undergo review by the Clean Air Scientific Advisory Committee (CASAC), an independent science advisory committee whose existence, review, and advisory functions are mandated by the Clean Air Act. The draft ISAs are subject to a public comment period before the final document is issued. The Risk and Exposure Assessment (REA) draws upon information and conclusions presented in the ISA to develop quantitative characterizations of exposures and associated risks to human health or the environment associated with recent air quality conditions and with air quality estimated to meet the current or alternative standard(s) under consideration.

Policy considerations of the REA results are considered in a Policy Assessment (PA), which is intended to bridge the gap between scientific evidence and technical information and the judgments required of the EPA administrator. Taking into consideration all the aforementioned assessments and reviews, the EPA develops and publishes a notice of proposed rulemaking. A public comment period, during which public hearings are generally held, follows, and after considering comments received on the proposed rule, the EPA issues a final rule.

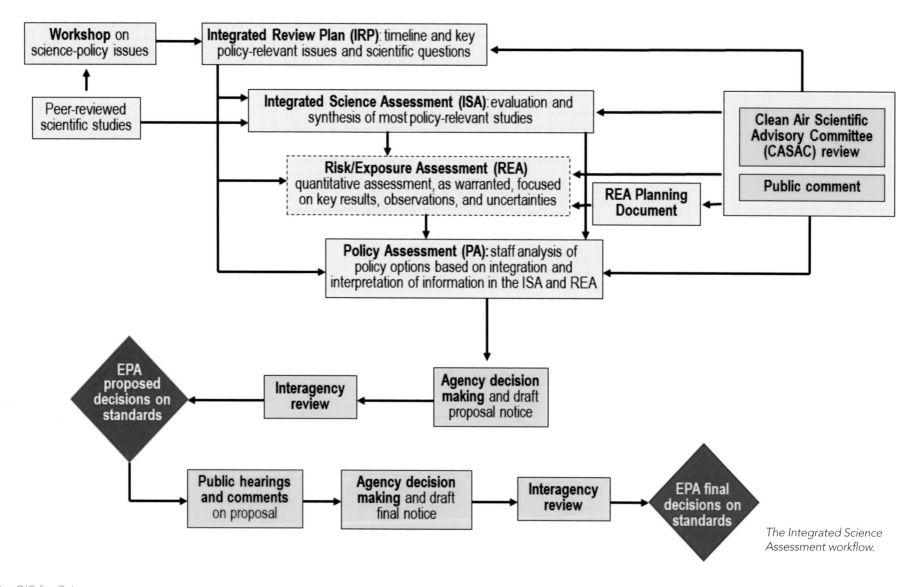

The Integrated Science Assessment workflow.

DATA ANALYSIS

A broad range of analyses can be undertaken to support the technical information these decisions are based on. Daily ozone (8-hour maximum) and fine particulate air (24-hour average) monitoring data from the National Air Monitoring Stations/State and Local Air Monitoring Stations (NAMS/SLAMS) are examples of the types of data used to inform decision-makers. These data provide on-the-ground measurements of the quality of the air we breathe every day. Since monitors are concentrated in populated areas and are more limited in the rural parts of the country, data fusion methods can be used to better understand air quality in areas without monitors.

Complex data fusion techniques allow scientists to incorporate information from multiple sources to come up with a better product than individual sources of data provide on their own. Because many factors drive air quality, predicting local air quality conditions in areas without monitors depends on identifying appropriate spatial modeling techniques and the right covariates. Some of the most advanced and recent

techniques, such as the empirical Bayesian kriging (EBK) regression prediction method, promise more accurate predictions than the other spatial interpolation models. Another advanced method developed by the EPA is named the Downscaler Model. This model fuses outputs from a gridded atmospheric model known as the Community Multiscale Air Quality Model (CMAQ) with point air pollution measurements from air quality monitors to produce an improved air quality surface for the entire country, which is extremely beneficial to the decision-making process. This method uses probability to represent the uncertainty of the input parameters and the uncertainty of the output.

Recent developments and improvements in GIS and spatial modeling allow the EPA to more accurately depict air quality nationwide. Additionally, the Geospatial Platform (GeoPlatform) is democratizing access to many data sources. Easy access to a suite of tools, apps, and data enables scientists to continually improve air quality spatial prediction models.

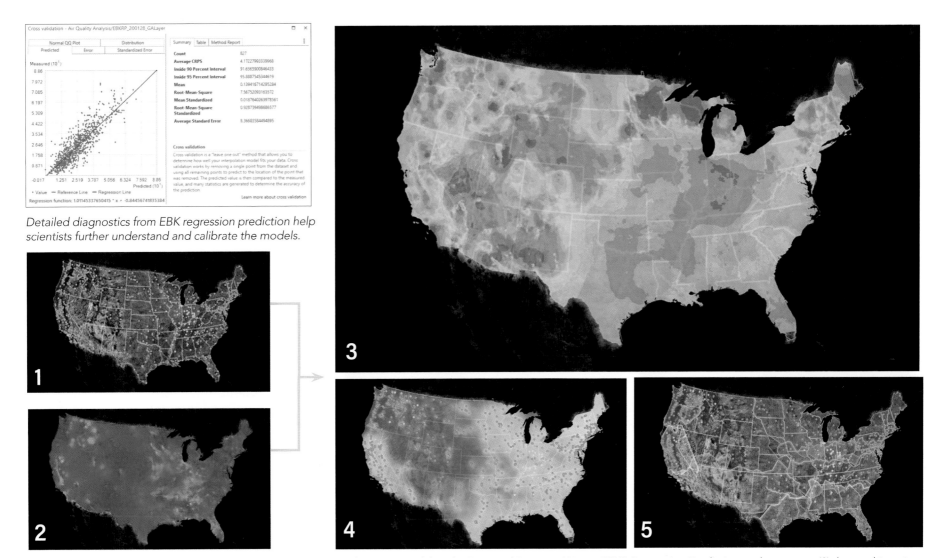

Detailed diagnostics from EBK regression prediction help scientists further understand and calibrate the models.

Using (1) CMAQ and (2) proximity to the US road network as explanatory variables, the Empirical Bayesian Kriging (EBK) Regression Prediction tool generates (3) the resulting air quality surface output. Supplemental views of the (4) prediction standard error and (5) subset polygons corresponding to regional models demonstrate the advantages of EBK regression prediction compared to other approaches. Local conditions can affect air quality in many ways, allowing the model to tune itself in different areas and account for local effect and providing better results when compared with fitting a global model.

IMPLEMENTING AIR QUALITY STANDARDS

Within two years of setting a new or revised standard, the EPA designates areas as meeting (attainment) or not meeting (nonattainment) the standard. Final designations are based on air quality monitoring data, state or tribal government recommendations, and technical information.

The EPA has become more innovative in recent years with the designation process, using more advanced spatial analysis to inform the recommended attainment and nonattainment areas, including improvements in spatial interpolation or the estimation of observation in places where data have not yet been collected.

Innovative web applications and analyses provide a wide range of data to internal and external stakeholders to inform the boundaries of nonattainment areas. They can help decision-makers understand how emissions, meteorology, and geography may affect their local air pollution concentrations. These boundaries are important because they indicate the area where air quality is above the standard and help state and local decision-makers, and tribal leaders pinpoint the pollution emission sources contributing to poor air quality, and to implement programs and control measures to improve air quality in their nonattainment areas.

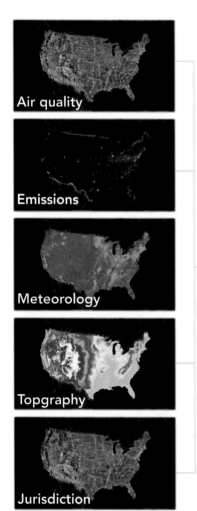

Air quality

Emissions

Meteorology

Topgraphy

Jurisdiction

Designation boundaries

The EPA's Ozone Designations Mapping Tool informs nonattainment area boundaries by combining data from five layers into a synthsized map of designation boundaries. Through GeoPlatform, EPA creates short-lived and limited-use designer apps like this to help policy makers and the public interactively deliberate alternative policy scenarios. The EPA creates these web applications to address specific problems and replace static figures, maps, and data.

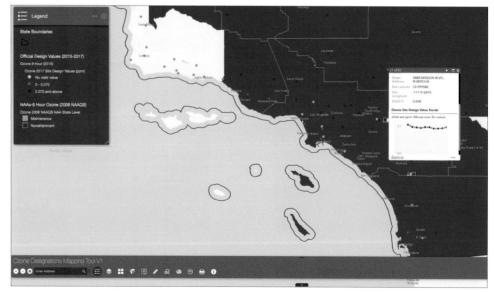

Rich data at the proposed designation boundary level comes to life in the EPA's Ozone Designations Mapping Tool.

MONITORING AIR QUALITY

Air quality data are spatial by definition: sources of pollution exist at specific locations. A network of thousands of outdoor air quality monitoring stations across the United States, Puerto Rico, and the US Virgin Islands is critical for monitoring changes in air quality, particularly important in areas of nonattainment. The data are compiled into the Air Quality System (AQS) database and available to the public via the AQS application program interface (API), in aggregated form via the EPA AirData website, and as dynamic GIS services via the ArcGIS Living Atlas of the World®. These data assist a wide range of people, from the concerned citizen who wants to be aware of unhealthy air quality days in their region during a time period to regulatory, academic, and health research communities that need raw data for air quality studies.

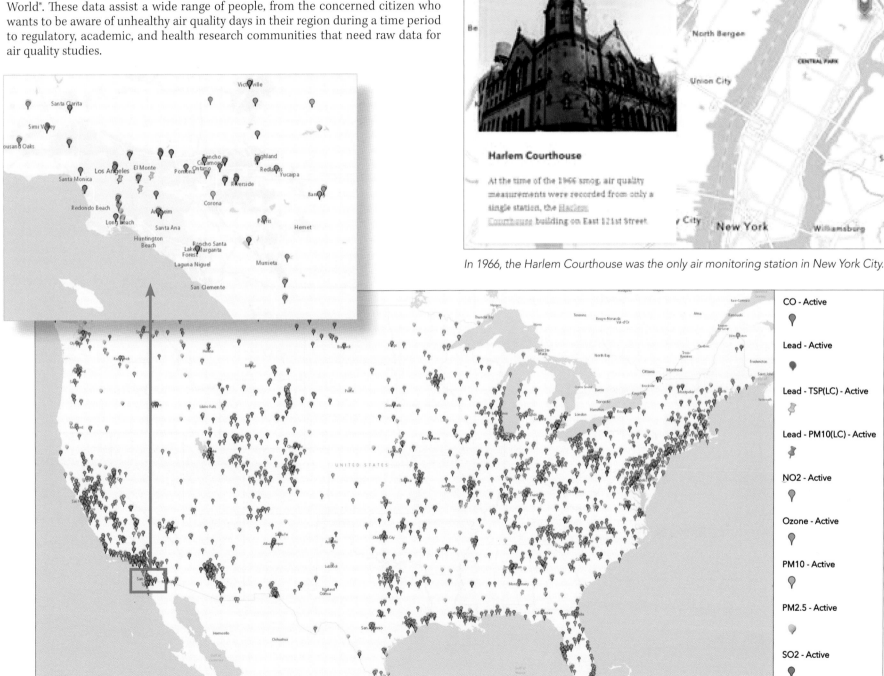

Harlem Courthouse

At the time of the 1966 smog, air quality measurements were recorded from only a single station, the Harlem Courthouse building on East 121st Street.

In 1966, the Harlem Courthouse was the only air monitoring station in New York City.

CO - Active

Lead - Active

Lead - TSP(LC) - Active

Lead - PM10(LC) - Active

NO2 - Active

Ozone - Active

PM10 - Active

PM2.5 - Active

SO2 - Active

Today, the EPA partners with local, state, and tribal agencies to manage a network of thousands of outdoor air quality monitoring stations.

TRENDS IN AIR QUALITY AND POLLUTION

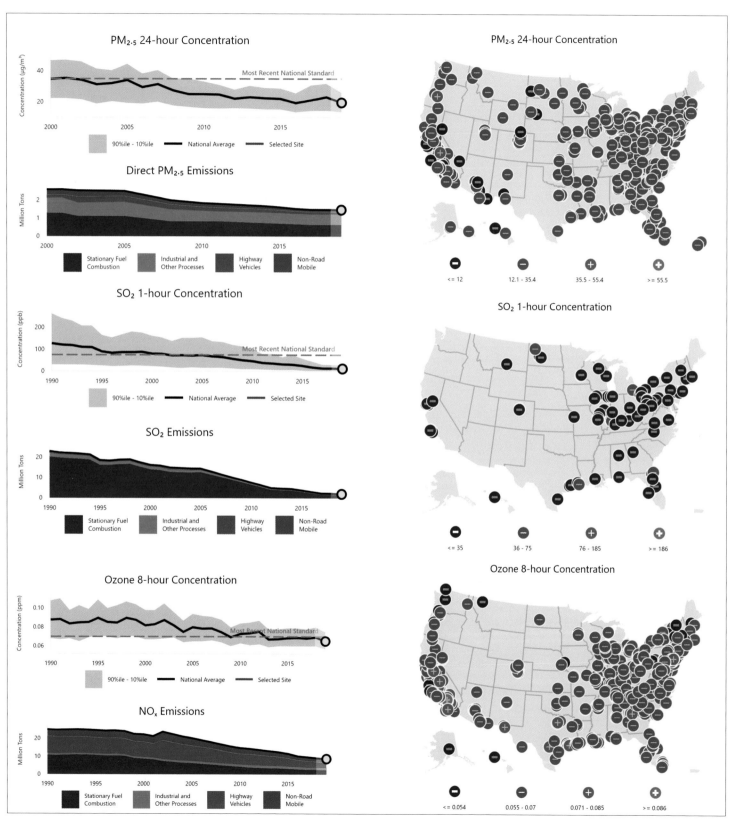

The Clean Air Act has provided public health protection since it became law in 1970. To highlight the achievements made in the United States, the EPA maintains interactive applications and map-driven reports outlining the trends in air quality.

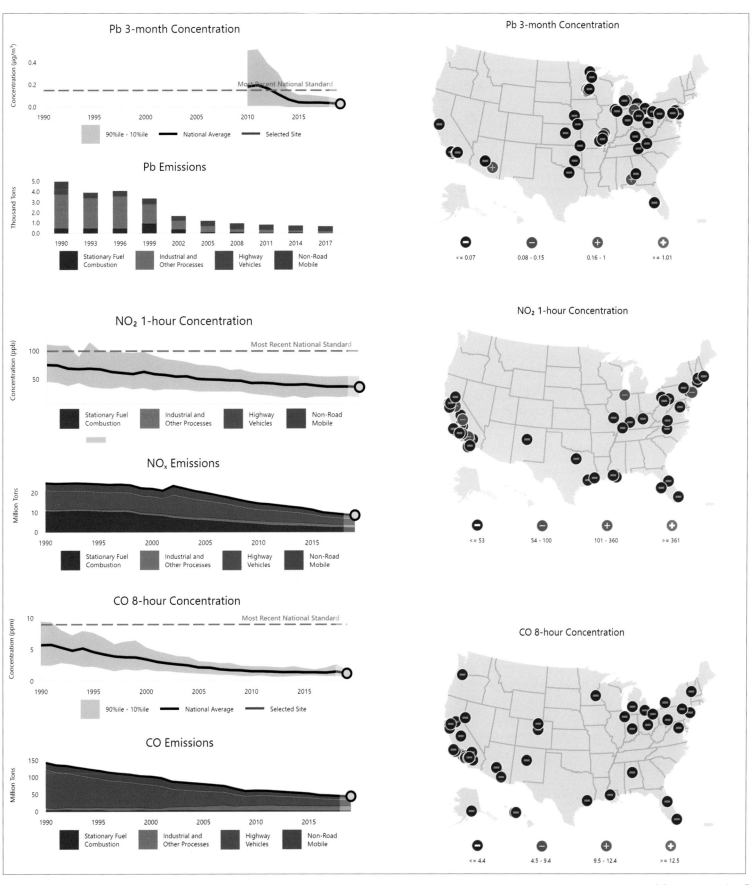

Through successful state-led implementation, numerous areas across the country are showing improvement, and fewer areas are in nonattainment. Since 2010, there were no violations of the standards for CO and NO_2.

INFORMING THE PUBLIC

EPA scientists aggregate, manage, and analyze raw data from thousands of outdoor air quality monitoring stations for policy-making purposes. The EPA also started the AirNow program to provide actionable information to the public for its day-to-day decisions. To better engage with the public, the EPA summarizes air quality information into the Air Quality Index (AQI), which explains how clean or polluted air is and what associated health effects might be of concern at their location. The AQI focuses on health effects that a person may experience.

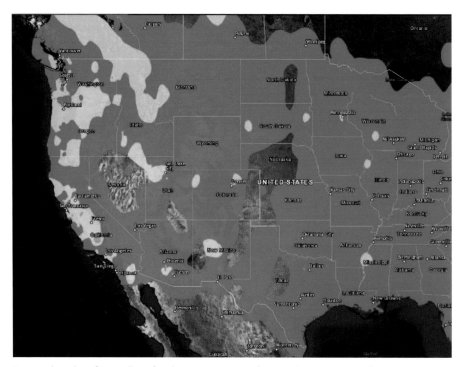

Interpolated surfaces allow for the assessment of air quality in areas without monitors. Currently AirNow uses the inverse-distance weighted (IDW) interpolation method because it provides fast results. The IDW method assumes that sites that are close to one another are more alike than sites that are farther apart. AQI values in areas without monitors are calculated using a weighted average of the values available at surrounding sites. With recent advances in algorithms and processing power, the AirNow team is exploring the use of more advanced and accurate but computationally taxing spatial interpolation methods. One method under consideration is the empirical Bayesian kriging regression prediction.

Air Quality Index Levels of Health Concern	Numerical Value	Meaning
Good	0 to 50	Air quality is considered satisfactory, and air pollution poses little or no risk.
Moderate	51 to 100	Air quality is acceptable; however, for some pollutants there may be a moderate health concern for a very small number of people who are unusually sensitive to air pollution.
Unhealthy for Sensitive Groups	101 to 150	Members of sensitive groups may experience health effects. The general public is not likely to be affected.
Unhealthy	151 to 200	Everyone may begin to experience health effects; members of sensitive groups may experience more serious health effects.
Very Unhealthy	201 to 300	Health alert: everyone may experience more serious health effects.
Hazardous	301 to 500	Health warnings of emergency conditions. The entire population is more likely to be affected.

AQI colors: The EPA has assigned a specific color to each AQI category so people can quickly understand whether air pollution is reaching unhealthy levels in their communities. The network of air quality monitoring stations is extensive, but spatial analysis is necessary to assess air quality for each person's location in the United States. Interpolated surfaces are therefore generated using each station's readings, allowing the public to receive an estimate of air quality even in regions with sparse air quality monitoring coverage.

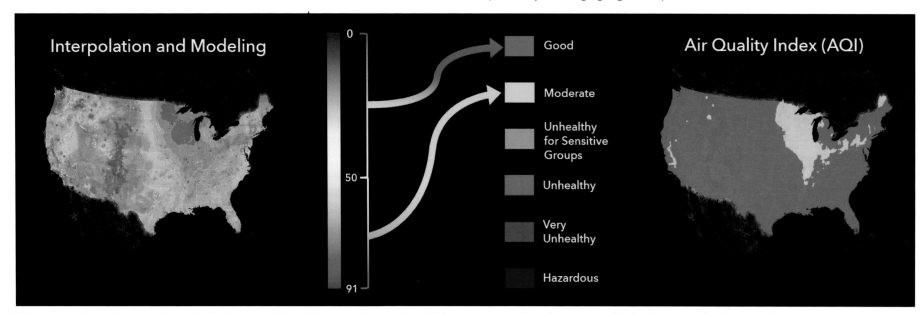

These maps show the air quality index for the entire United States on January 31, 2020. Largest area of moderate air quality for this particular day is in the Great Lakes region.

AIR QUALITY IS LOCAL

In March 2020, the EPA launched a new, locally focused AirNow website. This site provides the user with their local air quality information using their zip code or location from their phone. The public can immediately assess the air quality around them, learn more about historical air quality in their area, and the forecasted air quality for the following day. This information helps users plan their outdoor activities.

AirNow provides a window on local air quality data. Millions of people live in areas where air pollution can cause serious health problems and affect our daily lives. Like the weather, it can change from day to day. The AirNow program provides effective and modern ways to inform the public about air quality via the EPA AirNow website, widgets, and maps providing the current state of air quality around the nation.

Accurate prediction of local air quality relies on the fidelity of the input data that precisely depict local conditions in many ways. Human activity, topography, meteorology, and atmospheric conditions can affect local air quality in many ways.

Also new is the national Interactive Map of Air Quality, which provides additional information about the air quality monitoring locations, current AQI values, and pollutant concentrations.

In the 1970s Riverside, California, had some of the worst air quality in the nation because of its proximity to car-crowded Los Angeles. Today, the city routinely enjoys clear and blue skies but sometimes still experiences moderate air pollution, often in the form of particulate matter.

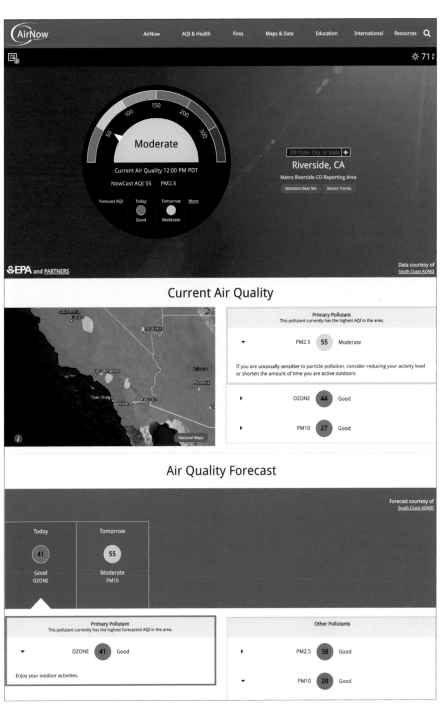

AirNow forecast for Riverside, California, captured in May 2020.

VISIBILITY STUDIES IN NATIONAL PARKS

One of the most basic forms of air pollution—haze—degrades visibility in many American cities and scenic areas. Haze results from sunlight encountering tiny pollution particles in the air, which reduce the clarity and color of what we see, especially during humid conditions. This pollution comes from a variety of natural and human-made sources. Natural sources can include windblown dust and soot from wildfires. Human-made sources can include motor vehicles, electric utility and industrial fuel burning, and manufacturing operations. The same pollution that causes haze also poses human and ecosystem health risks.

Since 1988, the federal government has monitored visibility in national parks and wilderness areas. A network of air quality monitors established in these treasured areas is providing a steady stream of data to assess the progress over time. Visibility has been improving due to pollution reductions resulting from many Clean Air Act programs, including the Regional Haze program, designed to help states make gradual progress to reach natural visibility conditions in national parks and wilderness areas. Thanks to the effectiveness of these programs, visibility in our national parks and wilderness areas is improving.

In general, in eastern parks and wilderness areas, the average visual range (the distance a visitor can see) has improved from 50 miles in 2000 to 70 miles in 2015. In western parks and wilderness areas, the average visual range has improved from 90 miles to 120 miles over the same period. The Regional Haze Storymap shows the visibility improvements in selected national parks. GIS is playing an important role in managing and disseminating these important data. Spatial analytics is helping to evaluate national and local visibility trends.

On a good day in Glacier National Park, visibility can extend more than 100 miles. The images were taken on different days at 3 p.m.

On a bad day, the visibility range can be just 12 miles or less.

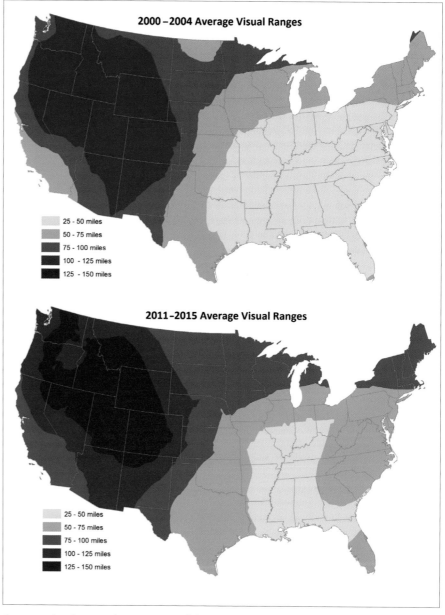

Average visual ranges have improved during the past two decades.

ECONOMIC GROWTH WITH CLEANER AIR

Between 1970 and 2019, the combined emissions of the six common pollutants dropped by 77 percent. This progress occurred while the US economy continued to grow, Americans drove more miles, and population and energy use increased. We don't need to compromise on air quality. It is important to get data and information into the hands of the public and decision-makers, which can be facilitated by interactive graphics, tools, and story maps. Together, we can make decisions that benefit the economy and help ensure we continue to improve the air for current and future generations.

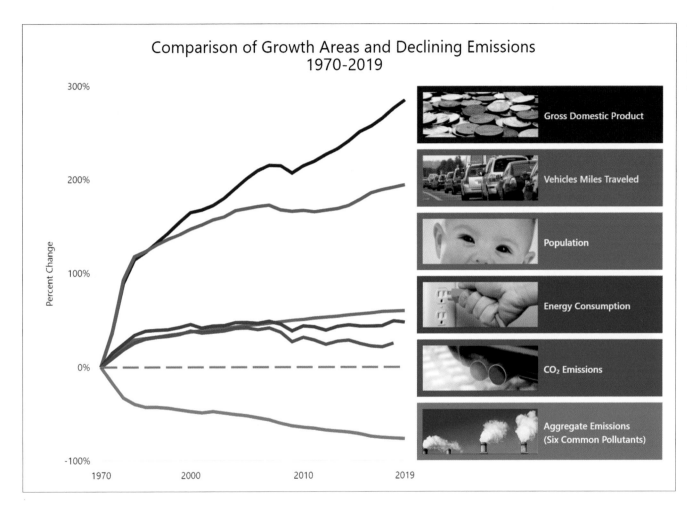

The data clearly show that the US can experience economic growth while reducing air pollution emissions and improving conditions for its citizens and the planet.

NOTES

Bell, M.L., Davis, D.L., and Fletcher, T. 2004. "A Retrospective Assessment of Mortality from the London Smog Episode of 1952: The Role of Influenza and Pollution." *Environmental Health Perspectives* 112 (1, January): 6–8. doi:10.1289/ehp.6539. PMC 1241789. PMID 14698923.

Gribov, A. and Krivoruchko, K., 2020. "Empirical Bayesian Kriging Implementation and Usage." *Science of the Total Environment* 722. doi: 10.1016/j.scitotenv.2020.137290.

Krivoruchko, K. and Alexander Gribov, A. 2019. "Evaluation of Empirical Bayesian Kriging." *Spatial Statistics* 32. doi: 10.1016/j.spasta.2019.100368.

Wonnacott, T. H., and R. J. Wonnacott. 1990. *Introductory Statistics* (New York: John Wiley & Sons).

Graphic on page 130: EPA Clean Air Scientific Advisory Committee (CASAC), *Particulate Matter Integrated Review Plan for National Ambient Air Quality Standards (NAAQS)*, 2016.

Links to EPA Air Data and Air Trends websites can be found at GISforScience.com.

THE URBAN TREE CANOPY

A city's treescape (or lack thereof) has a big impact on the quality of life. A unique government and academic partnership uses lidar and GIS technology to help communities map, assess, and monitor their urban tree canopy.

By Jarlath O'Neil-Dunne, **University of Vermont;** Dexter Locke and J. Morgan Grove, **US Forest Service;** and Michael Galvin, **SavATree**

Tree canopy lines the shores of the East River in New York City, with the Robert F. Kennedy Bridge connecting Astoria to Randall's Island. The lidar data shown here was the foundational dataset for New York City's most recent tree canopy assessment.

THE SUSTAINABLE CITY

The consolidation of the human population in urbanized areas during the Industrial Revolution of the late 1800s and early 1900s caused major challenges for cities across the United States. Raw sewage running down streets, air pollution, and contaminated water were just some of the issues that left urban residents in unhealthful living conditions. Over time, cities became more sanitary as they addressed these issues and improved their treatment plants, sewer and stormwater systems, and other facilities known as *gray infrastructure*. As a result, residents living in urban areas experienced measurable improvements in their quality of life. For the most part, US cities today are no longer focused on addressing sanitary ills, but they face a new set of challenges.

Today's environmental and livability challenges, such as climate change and urban heat islands, are in many ways more difficult to solve than those faced in the 19th century. Cities must become more sustainable and livable to meet these challenges, even as demographic shifts bring more people into urbanized city cores. Yet cities lack the kind of traditional gray infrastructure fixes when they look for ways to fight climate change or reduce the impacts of urban heat islands. And in a global marketplace, attracting the best companies and the brightest minds to a city is no longer assured simply because of its historically dominant presence in an economic sector. To remain competitive, a city must provide other amenities such as parks and tree-lined streets, which in turn requires revenue from the city's tax base.

To address these challenges, cities increasingly turn to green infrastructure to preserve and connect open spaces, watersheds, wildlife habitats, parks, and other natural landscapes. Growing and sustaining a strong tree canopy is a key strategy of green infrastructure. Trees benefit the ecosystem in many ways, from reducing peak summer temperatures to providing wildlife habitat to reducing stress in the human population. Cities have realized that they must manage trees as a crucial asset if they are to become sustainable, keep residents happy, and attract commercial entities.

A tree stands in front of row houses in the Reservoir Hill neighborhood of Baltimore. The city is a leader in mapping and protecting urban forests.

Knowing what you have so you can manage it

Baltimore, Maryland, like a lot of big US cities, has actively recognized the importance of its urban tree canopy in providing shade, ambiance, and character, not to mention the climate-cooling and carbon-capture benefits. In 2004, city officials approached a team that included US Forest Service scientists and academic researchers, with two simple questions:

1. How much of the city is covered by tree canopy?
2. How much land is available to plant new trees?

Baltimore, like many cities, had used GIS to map, manage, and monitor its infrastructure for years. Baltimore used digital parcel maps to collect taxes, updated gray infrastructure databases such as street centerlines and building footprints to stay current with changes, and relied on innovative online dashboards to report and analyze crime. The city had a lot of geospatial information at its fingertips but lacked information about green infrastructure when it decided to set a realistic long-term goal for tree canopy coverage. Elected leaders realized they could not set a goal for tree canopy without knowing how much tree canopy they had. They also needed to know the amount of available land available for new tree canopy.

Cities in general lacked information about their tree canopy partly because they did not traditionally view trees as a crucial asset. Only fairly recently have they understood and accepted the value of green infrastructure in urban areas. From a GIS perspective, technological reasons also prevented strong analysis of tree canopy until recently. First, cartographers historically mapped land cover at a resolution of 30 meters. While mapping land cover at that resolution works well to examine broad areas, 30-meter pixels are too coarse to analyze tree canopy in urban areas. A single pixel might contain dozens of land cover features and cross multiple property boundaries. Cities have had access to high-resolution imagery of urban areas for many years. However, they acquired the data under leaf-off conditions, which supported gray infrastructure and property parcel mapping but served as a poor source for assessing tree canopy. In addition, building shadows made it difficult to use overhead imagery to map trees in major cities, regardless of the spatial resolution.

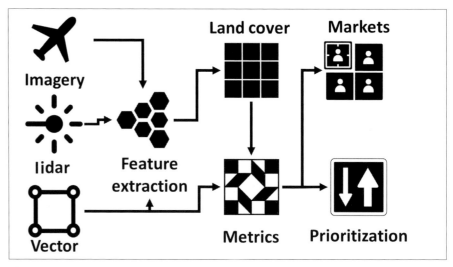

The Urban Tree Canopy Assessment Protocols form a comprehensive analytical framework for providing resource managers with insights into their urban forests.

MAPPING THE URBAN FOREST FROM ABOVE

The early 2000s brought a new era of remote sensing that made it possible to efficiently and effectively map tree canopy in urban areas. High-resolution digital imagery and lidar became increasingly prevalent, while new object-based approaches to feature extraction provided a way to automate land cover mapping. Commercial satellites offered detailed, on-demand acquisition, and the National Agriculture Imagery Program (NAIP), founded to support agricultural mapping, acquired imagery of entire states during the growing season while trees had their leaves. The combined result of these technologies allowed our team to develop workflows to map the tree canopy and other land cover classes that helped decision-makers estimate available land to plant trees.

Despite the technological advances, urban land cover mapping remained challenging. Cities are heterogeneous in the horizontal and vertical planes. The morphology of trees differs, depending on whether they grow without competition on the street or as part of a patch in an urban forest. Building shadows can obscure trees, and utility poles can look like trees in lidar data. The resolution of the remotely sensed data, while critical for mapping fine-scale urban features, meant that a land cover dataset for a single city could be made up of tens of billions of pixels, eclipsing the size of 30-meter land cover datasets for the entire United States.

The team's approach to high-resolution land cover mapping has centered on the implementation of object-based feature extraction techniques. Objects have the advantage in that they can contain information from raster, vector, and point cloud datasets. Objects are also spatially aware, in that they have inherent information on other objects, such as the relative border. Through an iterative process that employs segmentation, classification, and morphology algorithms, we can minimize the limitations of the input data and maximize their strengths. The land cover mapping workflow starts by classifying objects based on basic properties (e.g., height and tone).

As the workflow progresses, increasing amounts of spatial information are used in the classification process, such as relative border and distance. This approach allows us to eliminate almost entirely inconsistencies in the source data. For example, offsets between the lidar and the imagery due to building lean may make a building edge object have the height and spectral properties of tree canopy, but its spatial properties—its length-to-width ratio and relative border to the building—can be used to classify it correctly. Morphology algorithms help to improve the cartographic appearance of the land cover features, squaring up buildings and smoothing the edges of tree canopy. Automation itself is never enough, and good old-fashioned manual heads-up digitizing still serves as the final check on all our land cover maps.

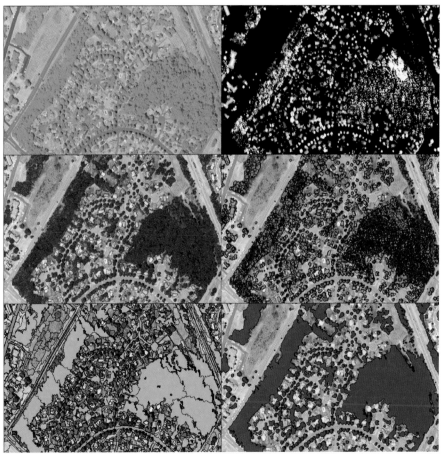

Feature extraction workflow showing the progression from source data (top), to segments and initial classification (middle), to contextual information and final classification (bottom).

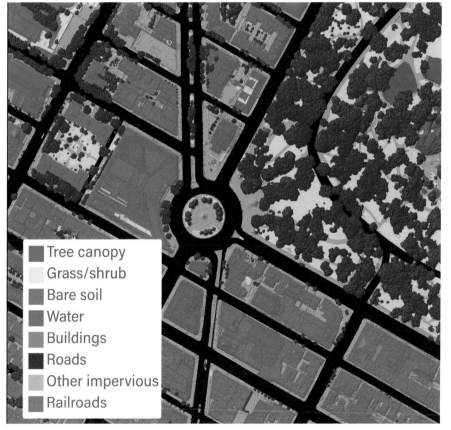

Tree canopy
Grass/shrub
Bare soil
Water
Buildings
Roads
Other impervious
Railroads

High-resolution land cover mapping is the foundation of an urban tree canopy assessment. This example is from the 2017 mapping of New York City.

DECISION SUPPORT TOOLS

Assessment

Land cover mapping proved to be valuable but alone did not provide enough information for cities to make informed decisions about their tree canopy. The collective benefits of tree canopy can be geographically broad, but the actual ownership of the tree canopy varies because most city land is privately owned. Cities also must understand their tree canopy at multiple geographical levels. A school district seeking insights into the percentage of tree canopy at each school needs parcel-level data. Elected city councilors want to understand the tree canopy in their district. Managing water quality requires the summation of tree canopy information by watershed.

An urban tree canopy (UTC) assessment summarizes land cover data within geographical units to calculate the existing tree canopy and possible tree canopy. Existing tree canopy is the tree canopy the community has right now. Possible tree canopy is considered the area without roads, buildings, water, or trees. These places could hypothetically support tree canopy. The purpose of defining the existing and the possible tree canopy is so that communities can understand what they have now as well as where the opportunities lie for establishing new tree canopy. The geographical units used to summarize tree canopy vary by community but generally include property parcels, political/administrative boundaries (e.g., council districts and neighborhoods), US Census block groups, and watersheds. UTC developed an ArcGIS geoprocessing model to compute the existing and possible tree canopy by geographical unit, providing an efficient means to batch process the data to compute the UTC metrics for large collections of data.

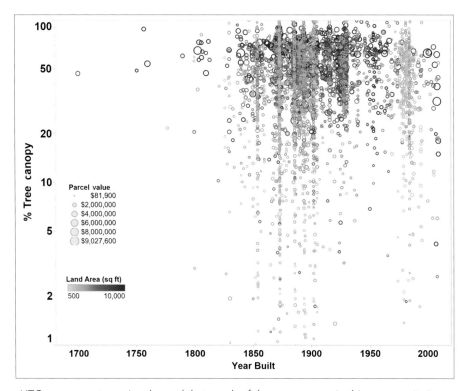

UTC assessment metrics showed that much of the tree canopy in this community is clustered in parcels that are 80 to 120 years old. The trees planted when these homes were built may now be reaching their life expectancy; thus, homeowners should plant new trees now to maintain current tree canopy coverage.

The metrics in the assessment phase provide communities with the information they need to set a tree canopy goal. This integration of data also offers insights into the relationships between tree canopy and other variables along with cross-tabulation information. A classic environmental justice example is to examine the relationship between wealth and tree canopy in public rights-of-way. This kind of study can help to determine whether wealthy residents, with greater means and political access, have disproportionately more street tree canopy. Another example is the founder's effect. Trees rarely survive new construction, and thus one of the first things a developer or homeowner does on a new property is plant trees. These trees will yield robust canopies many decades after the house is built. In areas where the houses were constructed at a similar time, tree canopy will exhibit a characteristic "rise and fall" as trees reach peak canopy at a similar time, followed by a sudden drop as the trees die. The integration of land cover with property parcel data and assessor's records can provide predictive analytics that allow cities to target landowners ahead of time, reminding them that they need to plant new trees now if they want to sustain their tree canopy in the long term.

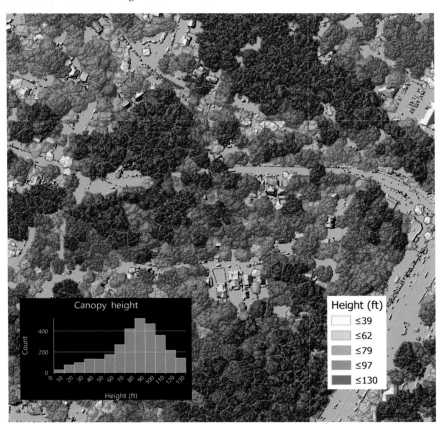

Canopy height information derived from lidar provides insight into the structure of a community's urban forest.

Prioritization

Often after a tree canopy goal is set, urban foresters, planners, and other decision-makers want to know how to reach their goal. UTC prioritizations first identify the places lacking the benefits of trees and then help identify organizations whose mission or mandate reflects that management priority. For example, trees reduce summer temperatures by blocking the sun and through evapotranspiration (exhaling water vapor). High temperatures in the summer can be lethal to humans. Public health officials may then choose to prioritize places to plant trees that are hot in the summer and where the young and old live—people most vulnerable to heat. UTC prioritization is a set of GIS tools and a stakeholder engagement process.

In the case of Baltimore, the team gathered members of 25 organizations whose missions or mandates could be achieved—in part—by increasing tree canopy. We assembled a "menu" of data reflecting different ecosystem services that trees provide, which could logically be linked to the participating organizations' mission or mandate. For example, by removing impervious surfaces and by planting trees, flooding may be reduced. Areas prone to flooding could also be identified by the point locations of service requests—Baltimore City's 311 nonemergency government hotline. Each organization could distribute 10 votes across all the items in the menu to best reflect their organization's priorities. Or they could place all 10 votes in the income box, which would indicate that they are only interested in planting in low-income areas.

Using ArcGIS, UTC created custom maps for each of the 25 participating organizations to reflect their mission or mandate, using the weightings that each participant provided. A 26th map was created by summing together all 25 organizations' maps. That map is now in the city's sustainability plan as the official urban tree canopy prioritization map. One important benefit to this approach is that each organization can compare its own map to the map in the sustainability plan and to other organizations' maps. In this way, people can see where they have the same high priorities, for similar or dissimilar reasons. Groups can also see how they contribute to the city's overall goal.

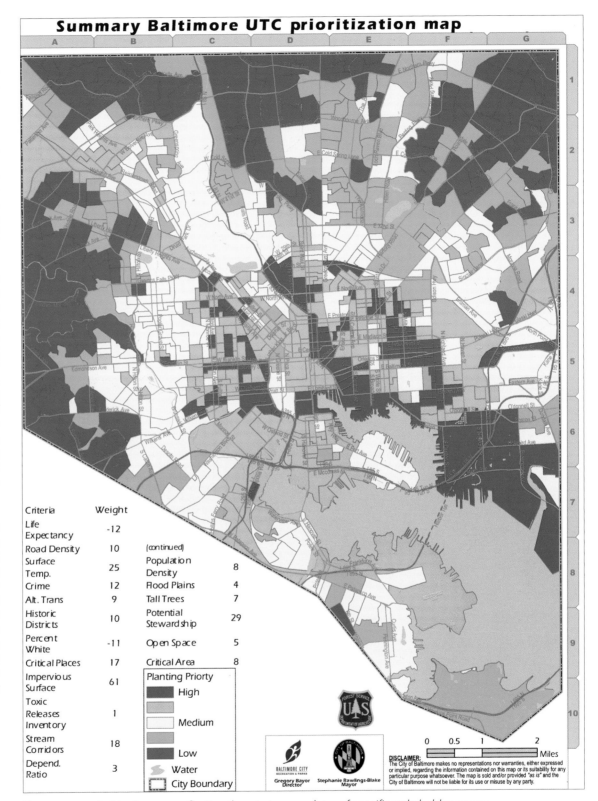

Summary Baltimore UTC prioritization map

Criteria	Weight
Life Expectancy	-12
Road Density	10
Surface Temp.	25
Crime	12
Alt. Trans	9
Historic Districts	10
Percent White	-11
Critical Places	17
Impervious Surface	61
Toxic Releases Inventory	1
Stream Corridors	18
Depend. Ratio	3

(continued)	
Population Density	8
Flood Plains	4
Tall Trees	7
Potential Stewardship	29
Open Space	5
Critical Area	8

Planting Priorty
- High
- Medium
- Low
- Water
- City Boundary

BALTIMORE CITY RECREATION & PARKS — Gregory Bayor Director

Stephanie Rawlings-Blake Mayor

FOREST SERVICE US DEPARTMENT OF AGRICULTURE

0   0.5   1   2 Miles

DISCLAIMER: The City of Baltimore makes no representations nor warranties, either expressed or implied, regarding the information contained on this map or its suitability for any particular purpose whatsoever. The map is sold and/or provided "as is" and the City of Baltimore will not be liable for its use or misuse by any party.

Tree canopy prioritization map reflecting the mission mandates of specific stakeholder groups.

THE NUMBER OF ACRES OF POSSIBLE TREE CANOPY

594 ACRES RESIDENTIAL   **383** COMMERCIAL   **324** INSTITUTIONAL   **163** RIGHTS-OF-WAY

"Residential has the most area for possible tree canopy."

UTC assessment infographic developed for decision-makers to help them understand who in the city has the most land available (termed possible tree canopy) for establishing new tree canopy.

Markets

UTC assessments provide baseline information often used for goal setting, and UTC prioritizations provide an implementation plan that identifies key areas for planting and common goals across organizations or institutions. However, residents across low- to high-priority neighborhoods have different motivations, capacities, and interests in urban and community forestry initiatives. A goal of UTC market analyses is to understand how participation in existing programs varies by geodemographic segment or market group. This data is typically used to understand purchasing behavior and to market consumer goods. However, previous research has shown that the amount of existing and possible tree canopy varies not just by household income but also by family structure such as marital status or number of children living at home.

Our market analyses help us compare where trees are being planted, through which planting program, who lives where the trees are planted, and how much tree canopy exists in that area. The results reveal how current programs are reaching—or not reaching—different social groups and how much tree canopy is currently available. The idea is to find out where alternative approaches could benefit additional tree planting based on the demographics and lifestyles of residents in different areas.

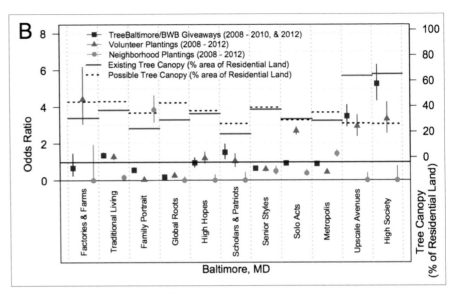

Odds ratios and 95 percent confidence intervals for urban greening programs on private residential land by Esri Tapestry LifeMode®.

Example of the land cover mapping carried out in support of San Diego's tree canopy assessment. More than 1 billion pixels of data were analyzed.

Monitoring

For cities that have completed an assessment, set a goal for tree canopy, prioritized areas for tree plantings, and implemented strategies to protect existing tree canopy and establish new tree canopy, the next logical step is to assess the change in tree canopy over time. Mapping tree canopy change is even more challenging than the initial land cover mapping. Estimating tree canopy change over time from remotely sensed data requires that the amount of change measured falls outside of the margin of error. For example, if the two tree canopy estimates, produced at different times, have a margin of error of +/− 2 percent, one cannot conclude that there is a 2 percent increase. The chief obstacle in tree canopy change is the source data.

The various imagery and lidar datasets used for monitoring tree canopy change are collected with other use cases in mind and different acquisition parameters, and then processed to different specifications. Even if individual tree canopy mapping done at two time periods was perfect, the process of differentiating the two tree canopy datasets would result in false change due to the issues mentioned earlier. To accommodate the challenges, we developed mapping protocols that minimize errors associated with mapping tree canopy change over time by mapping three categories at the tree scale—no change, loss, and gain. The assessment starts again with monitoring. The UTC used monitoring data to generate its assessment metrics. These assessment metrics provide information on changes at various geographical units. This information, in turn, is used to draw conclusions on driving factors.

Monitoring helps to reveal changes that can go unnoticed from the ground. UTC monitoring helped one community understand that despite its substantial investment in street tree maintenance and planning, tree canopy was declining in the community due to losses in residential backyards. These mostly unseen areas contained most of the city's tree canopy. In another city, monitoring revealed that neighborhoods in the lowest income quintiles lost the most tree canopy over a five-year period, even though they had the least tree canopy as a percentage of land area to begin with. This discovery had important environmental justice and land management implications. Monitoring does not always reveal bad news. A consistent finding is that existing tree canopy will continue to expand. Planting new trees can be expensive, and monitoring has helped cities understand that preserving what they have can result in greater gains at a lower cost per unit.

Gain
Loss
No change

Monitoring tree canopy change in New York City. The data on the left is overlaid on a lidar surface hillshade from the first time period, and the data on the right is overlaid on a lidar surface from the second time period. Purple represents no change in tree canopy, orange represents loss, and green represents gain.

UTC IN ACTION

What started as a pilot project in Baltimore, Maryland, has expanded to more than 80 communities in the United States and Canada. As of this writing, UTC assessments cover 8.5 million acres and include approximately 37 million people. UTC assessments are driving green decision-making in urban areas. Cities are setting tree canopy goals and building UTC mapping into their standard mapping updates alongside parcel and planimetric mapping. We are now in the era of UTC data informatics—the integration of UTC data with other citywide data used to measure the sustainability and resilience of neighborhoods and cities. This integration is possible because of *data hooks* associated with a parcel. These data hooks might include latitude and longitude for some data, such as maps of urban heat islands or flood zones, and an address for other data, such as crime, health, and water and energy use. These data hooks are the critical, unique connectors that make data interoperable and UTC data informatics possible. UTC data informatics and synthesis continue to increase in value because the amount of digital data about cities continues to grow at phenomenal rates. Data informatics and synthesis provide the basis for expanding how we understand the benefits and services of trees and canopy. In other words, because we can expand the types and number of environmental, social, economic, and health data that can be integrated with UTC data, we can employ a variety of techniques—hypothesis testing, machine learning, and time-series analysis—to gain novel insights into the effects that trees and canopy have on cities.

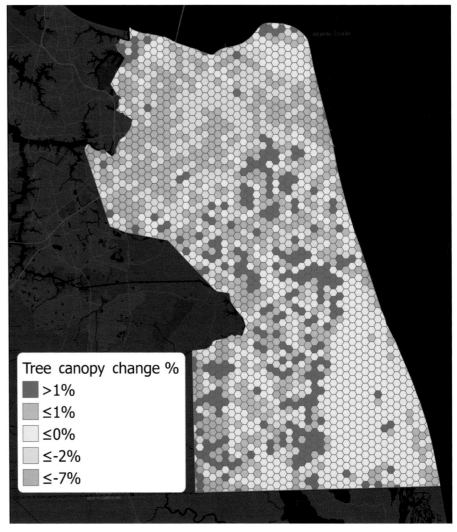

Tree canopy change summarized by 100-acre hexagons for the city of Virginia Beach for 2012–2018. Tree canopy decreased in the urbanized portions of the city to the north and increased in the natural areas, which tend to be in the central and southern portions of the city.

Resource managers can use the data from tree canopy assessments to find communities with similar tree canopy characteristics and share strategies for maintaining and increasing their urban tree canopy.

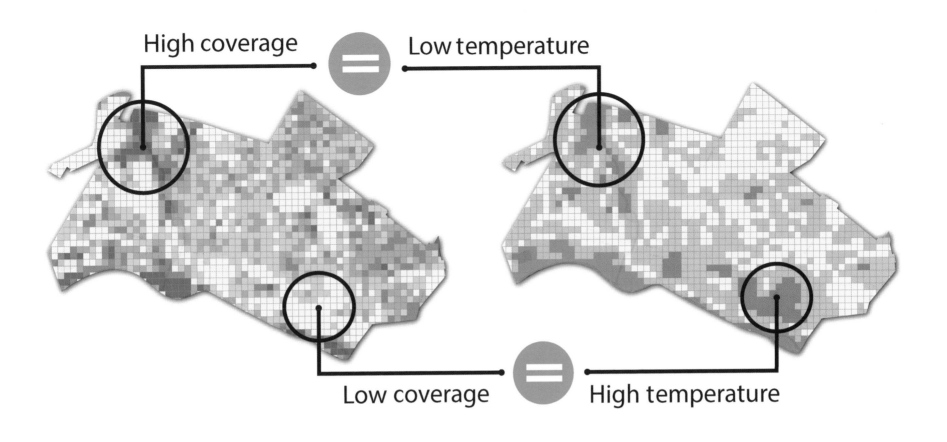

High coverage = Low temperature

Low coverage = High temperature

Tree canopy (%)

| 0% - 11% | 12% - 25% | 26% - 41% | 42% - 66% | 67% - 100% |

Surface temperature (Fahrenheit)

| 77 - 84 | 85 - 89 | 90 - 92 | 93 - 97 | 98 - 113 |

Trees provide important ecosystem services, such as reducing the urban heat island. Tree canopy mapping from high-resolution imagery combined with surface temperature mapping derived from Landsat satellite thermal imagery can illustrate the benefits that trees provide to a community.

Acknowledgments

We extend our gratitude to the US Forest Service and the dozens of communities throughout North America that collaborated with us on Urban Tree Canopy Assessments during the past decade.

MONITORING DISASTERS

Responding to major natural and human-caused hazards, the NASA Earth Applied Sciences Disasters Program collects, synthesizes, and shares data collected from dozens of Earth-orbiting satellites.

By Jeremy Kirkendall and Garrett Layne, **NASA Disasters Program**

Hurricane Florence—viewed here from the International Space Station on September 12, 2018—caused widespread flooding along the Carolina coast on the southeastern US seaboard.

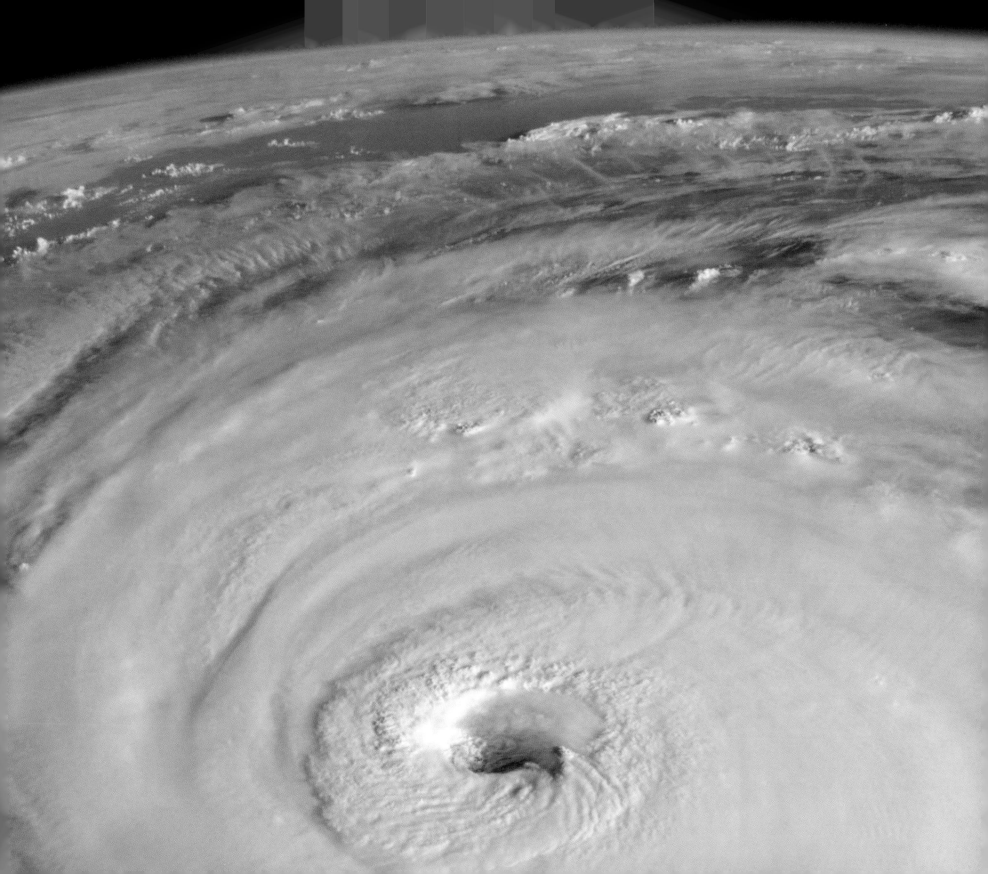

NASA EARTH SCIENCE

The NASA Disasters Program within Earth Applied Sciences is part of NASA's overall Earth Science Division (ESD), which is charged with delivering the technology, expertise, and global observations required to help scientists map the myriad of connections between Earth's vital processes and the effects of ongoing natural and human-caused changes.

Using observations from satellites, instruments on the International Space Station, airplanes, balloons, ships, and on land, ESD researchers collect data about the science of our planet's atmospheric motion and composition; land cover, land use, and vegetation; ocean currents, temperatures, and upper-ocean life; and ice on land and sea. These datasets, which cover even the most remote areas of Earth, are freely and openly available to anyone.

Of particular interest to the GIS and disasters mapping communities are the data coming from the Earth-orbiting satellites depicted here.

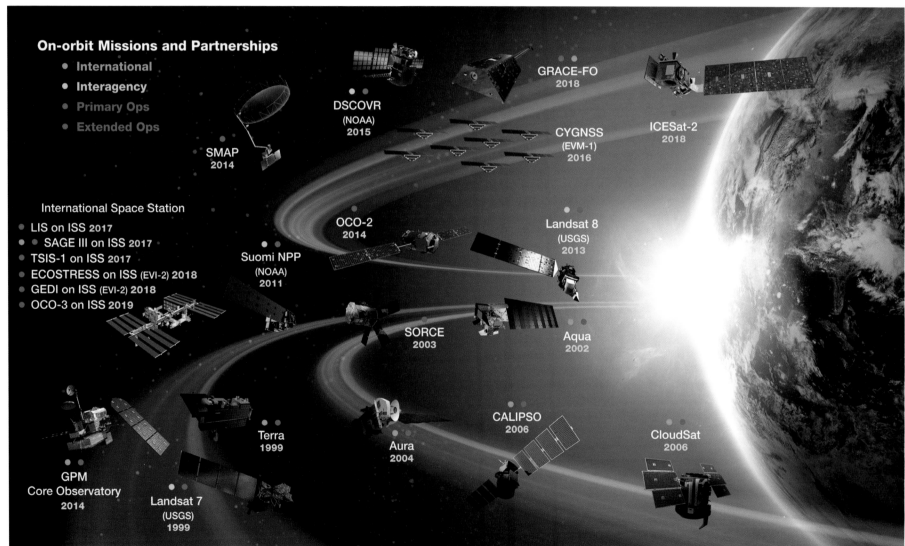

NASA's Earth Science group processes data from a fleet of orbiting satellites that continually image Earth across the complete electromagnetic spectrum. Mission-specific platforms like ERRA/AQUA and the MODIS instrument, plus a wide variety of emerging "microsatellites," produce data applicable to serious GIS analysis.

MISSION OF THE NASA DISASTERS PROGRAM

The NASA Disasters Program promotes the use of Earth observations for disaster management and risk reduction. The program coordinates data and information among its members to minimize the impact of disasters in collaboration with emergency management organizations, government officials, the private sector, humanitarian actors, and others. The program, through the development and contribution to actionable Earth science research, aims to enhance situational awareness and empower decision-making before, during, and after disasters. The program develops data products based on a solutions-oriented approach to identify vulnerabilities and assess risk factors to promote planning and mitigation, improve response, hasten recovery, and build resilience.

This holistic approach offers the ability to see the bigger picture and identify more valuable, expansive uses of data previously siloed by a specific research question or hazard type. NASA develops and combines data in new, innovative, and unique ways to fill gaps in information and support the needs of communities, properties, or economies deemed most vulnerable. Whenever possible, the program responds to direct requests for relevant data; opportunities to apply, advance, and evaluate disaster science and technology; potential transitions of science to operational users and collaborators; and advancement of science understanding and NASA capabilities.

The NASA Disasters Mapping Portal serves as a publicly accessible focal point to make usable, integrated, and visualized disaster GIS data available for further analysis by anyone. The portal provides context to data to demonstrate their value and potential individually and in combination with other products. Anyone can stream the products at no cost. The portal hosts event-specific products for disasters such as a tropical cyclone or earthquake, and it hosts near real-time products, many of which have global coverage.

One such product used to monitor hurricanes and other large-scale weather events is the Integrated Multi-satellitE Retrievals for Global Precipitation Measurement (IMERG)—essentially a measurement of precipitation accumulation. The following large-format composite depicts the situation during Hurricane Willa in 2018.

Accessible through disasters.nasa.gov, NASA Disasters Program serves as a virtual near real-time "news desk" providing a variety of information from across NASA's teams.

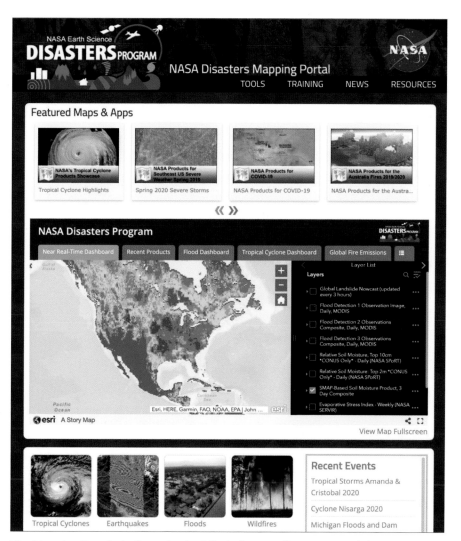

The Mapping Portal—built on the ArcGIS platform—offers a map and data-centric view of currently unfolding and historical disasters, accessible at maps.disasters.nasa.gov.

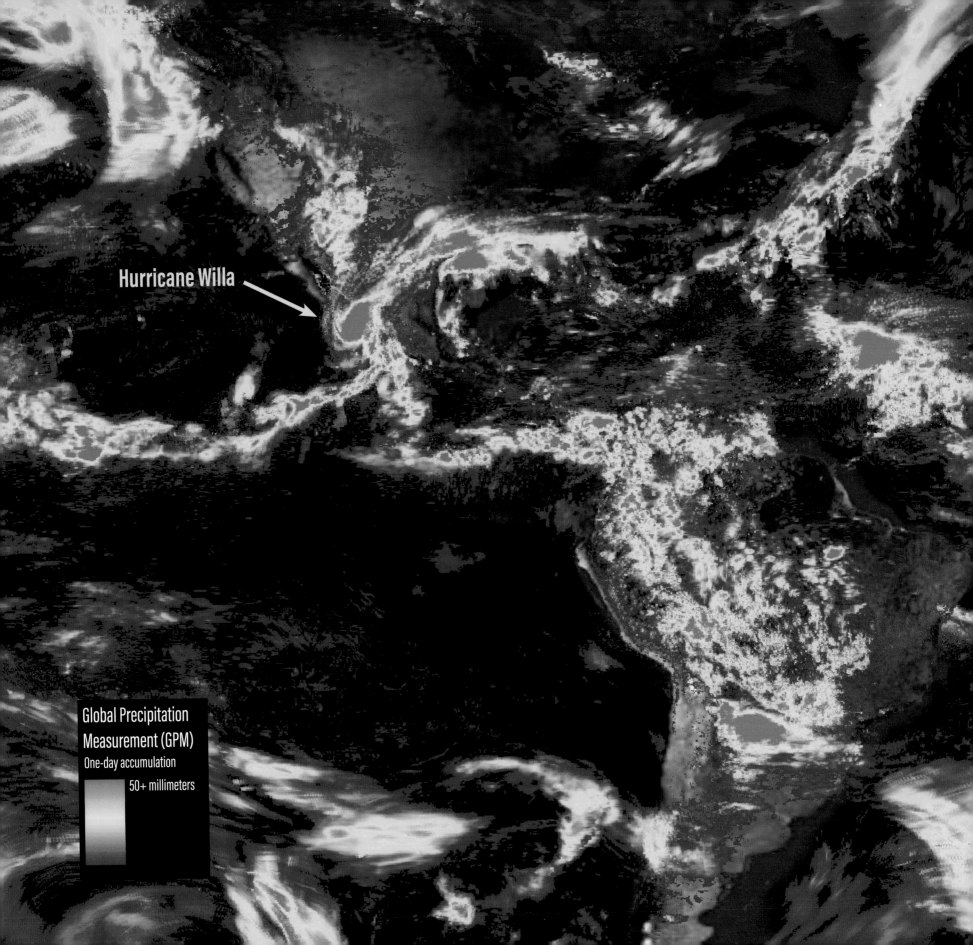

Hurricane Willa

Global Precipitation
Measurement (GPM)
One-day accumulation
50+ millimeters

GPM IMERG data showing one-day precipitation accumulation on October 23, 2018, during Hurricane Willa (off the coast of Mexico at the time this image was recorded).

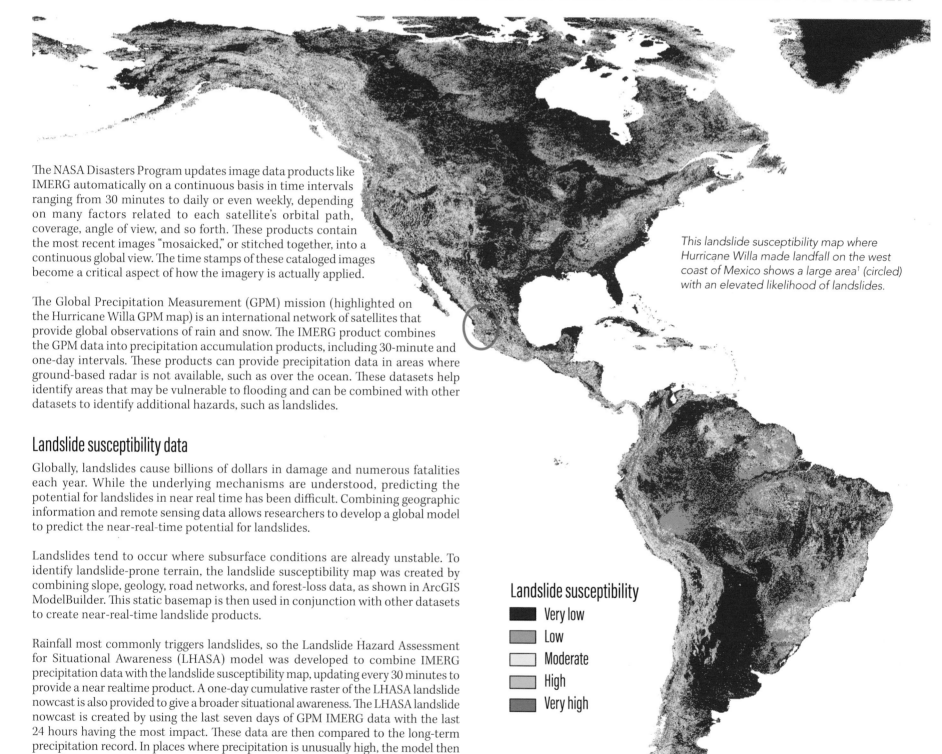

The NASA Disasters Program updates image data products like IMERG automatically on a continuous basis in time intervals ranging from 30 minutes to daily or even weekly, depending on many factors related to each satellite's orbital path, coverage, angle of view, and so forth. These products contain the most recent images "mosaicked," or stitched together, into a continuous global view. The time stamps of these cataloged images become a critical aspect of how the imagery is actually applied.

The Global Precipitation Measurement (GPM) mission (highlighted on the Hurricane Willa GPM map) is an international network of satellites that provide global observations of rain and snow. The IMERG product combines the GPM data into precipitation accumulation products, including 30-minute and one-day intervals. These products can provide precipitation data in areas where ground-based radar is not available, such as over the ocean. These datasets help identify areas that may be vulnerable to flooding and can be combined with other datasets to identify additional hazards, such as landslides.

This landslide susceptibility map where Hurricane Willa made landfall on the west coast of Mexico shows a large area[1] (circled) with an elevated likelihood of landslides.

Landslide susceptibility data

Globally, landslides cause billions of dollars in damage and numerous fatalities each year. While the underlying mechanisms are understood, predicting the potential for landslides in near real time has been difficult. Combining geographic information and remote sensing data allows researchers to develop a global model to predict the near-real-time potential for landslides.

Landslides tend to occur where subsurface conditions are already unstable. To identify landslide-prone terrain, the landslide susceptibility map was created by combining slope, geology, road networks, and forest-loss data, as shown in ArcGIS ModelBuilder. This static basemap is then used in conjunction with other datasets to create near-real-time landslide products.

Rainfall most commonly triggers landslides, so the Landslide Hazard Assessment for Situational Awareness (LHASA) model was developed to combine IMERG precipitation data with the landslide susceptibility map, updating every 30 minutes to provide a near realtime product. A one-day cumulative raster of the LHASA landslide nowcast is also provided to give a broader situational awareness. The LHASA landslide nowcast is created by using the last seven days of GPM IMERG data with the last 24 hours having the most impact. These data are then compared to the long-term precipitation record. In places where precipitation is unusually high, the model then uses the LHASA susceptibility map to determine whether the area is vulnerable to landslides. If the area is vulnerable, the model produces a nowcast identifying the area as having a high or moderate likelihood of landslide activity.

Landslide susceptibility
- Very low
- Low
- Moderate
- High
- Very high

When Category 3 Hurricane Willa approached the Mexican state of Sinaloa on October 23, 2018, as much as 15 inches of rain fell across the region, causing the nowcast to show a widespread risk of landslide. When comparing the one-day precipitation accumulation to the landslide susceptibility map, the areas of overlap between the heaviest precipitation and highest landslide probability result in the landslide nowcast.

The nowcast system has been evaluated by comparing the nowcasts to each of 3,989 landslide event points in the Global Landslide Catalog. Since most landslides occur in places with no observations, it was not possible to verify that the global LHASA nowcast is accurate in all locations. However, it does provide a near-real-time global summary of landslide hazards that may be useful for disaster response agencies, international aid organizations, and others who would benefit from situational awareness of potential landslides in near real time.

In the future, the LHASA model will be evaluated using the Cooperative Open Online Landslide Repository (COOLR), which combines data from the Global Landslide Catalog and data from citizen scientists in an effort to reduce inconsistencies in how landslides are reported in different regions. Anyone can view landslides as well as report a landslide event using the Landslide Viewer and Landslide Reporter GIS web applications. Growing this global landslide database will help validate and improve LHASA, as well as enable the landslide community to advance landslide research and understanding of where and when landslides are occurring.

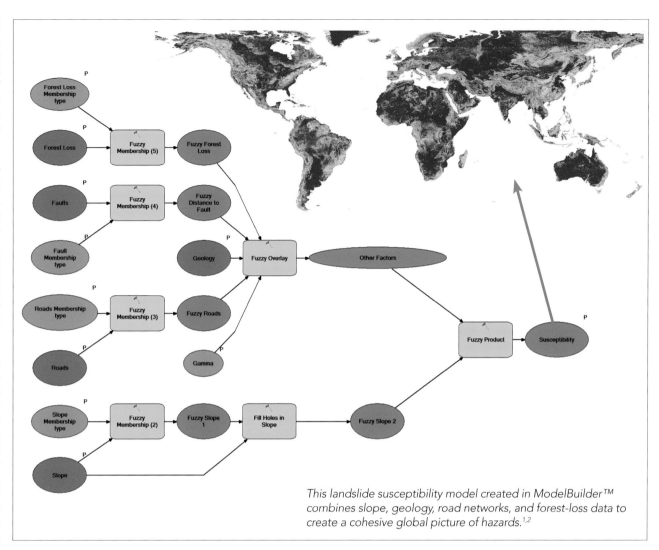

This landslide susceptibility model created in ModelBuilder™ combines slope, geology, road networks, and forest-loss data to create a cohesive global picture of hazards.[1,2]

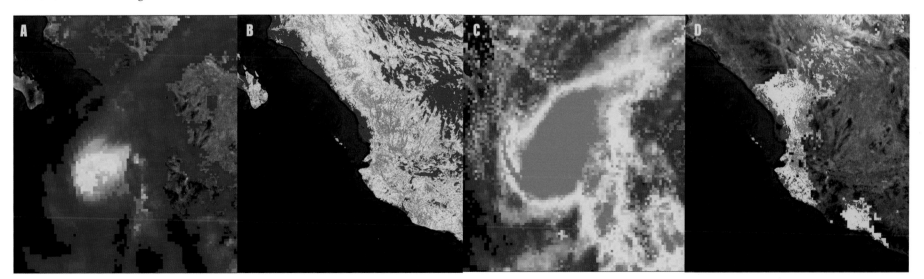

(A) IMERG 30-minute precipitation accumulation product for Hurricane Willa on October 23, 2018; (B) the Landslide Susceptibility Map of where Willa made landfall shows a large area with an elevated probability of landslide; (C) IMERG one-day's precipitation accumulation product for all of October 23; (D) the landslide one-day's nowcast for the date Willa made landfall shows areas of moderate and high landslide likelihood caused by the storm's heavy rainfall.[1,2]

HURRICANE DORIAN: OPTICAL AND SYNTHETIC APERTURE RADAR (SAR) COMPARISON

Category 5 Hurricane Dorian struck the northern Bahamas on September 1, 2019, and stalled there for more than 40 hours, causing heavy rain and catastrophic storm surge. By combining multiple days of GPM data, NASA created a total rainfall product to highlight hardest-hit areas, with some locations experiencing more than 4 feet of rain.

The National Oceanic and Atmospheric Administration's Geostationary Operational Environmental Satellite (GOES-East) satellite provided rapid, highly detailed imaging so that forecasters would have critical information about the storm's movement. However, the optical sensor could not detect flooding beneath the clouds while the storm moved slowly over the Bahamas.

In the evening hours of September 2, 2019, the Copernicus Sentinel-1 satellite also passed over the Bahamas, with a synthetic aperture radar (SAR) system capable of imaging the land surface through Dorian's clouds and rainfall. The SAR data, provided by the European Space Agency (ESA), were then processed into a Flood Proxy Map (FPM) by the Advanced Rapid Imaging and Analysis (ARIA) team at NASA's Jet Propulsion Laboratory in collaboration with the Earth Observatory of Singapore.

To help communicate the benefits of SAR during these cloudy conditions, the NASA Disasters Mapping Portal built an interactive web app to display the FPM layer beneath GOES-East imagery from the same time as the Sentinel-1 overpass. By moving the spyglass widget in the app, users can view flooded regions, shown in blue, as if they were peering beneath the clouds. The yellow box indicates the extent of the SAR data collected, showing the limitation of SAR's narrower but higher spatial-resolution swath. Thanks to SAR's cloud-penetrating capabilities, flooding could be assessed before the storm and associated clouds cleared, allowing emergency responders to act more quickly than if only relying on optical-based imagery.

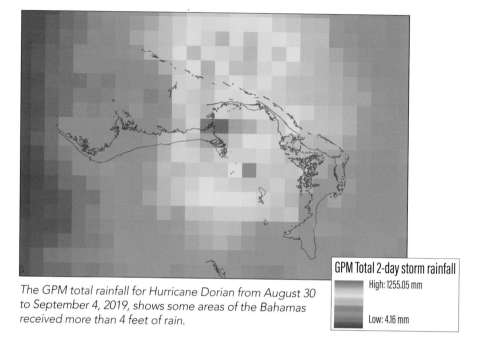

The GPM total rainfall for Hurricane Dorian from August 30 to September 4, 2019, shows some areas of the Bahamas received more than 4 feet of rain.

GPM Total 2-day storm rainfall
High: 1255.05 mm
Low: 4.16 mm

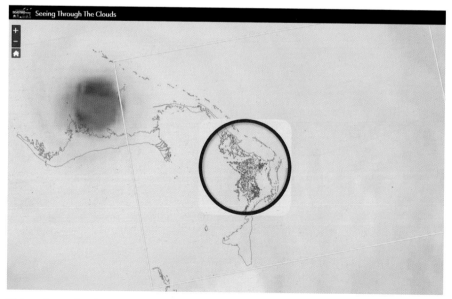

This web application uses a spyglass widget to let users see beneath the NOAA GOES-East optical imagery of Hurricane Dorian, revealing a Copernicus Sentinel-1 SAR-derived flood proxy map that identifies areas likely flooded by the storm. The optical imagery and proxy map shown here were captured within minutes of each other.

This flood proxy map image, captured on September 4, 2019, shows heavy flooding (in blue) at the Grand Bahama International Airport (near the top of the image) and throughout the city of Freeport.

3 miles

Taken on the morning of November 8, 2018, just a few hours after ignition, this Landsat 8 image shows how far the Camp Fire north of Sacramento, California, had spread by 10:45 a.m. By that evening, the fire had burned over 18,000 acres and remained zero percent contained. Shortwave infrared light highlights the active fire.

SOIL MOISTURE AND EVAPORATIVE STRESS INDEX

Before disaster strikes, satellites sometimes can observe warning signs to assess changes in risk. Monitoring soil moisture and evapotranspiration can help detect when and where vegetation dries out, which creates additional fuel for wildfires. The Camp Fire in November 2018 saw dry conditions for an extended period of time because of persistent high pressure over the western United States. The vegetation dried out in the weeks before the fire, helping to fuel the destructive blaze, which devastated the Northern California town of Paradise and surrounding rural communities, killing scores of people.

The Land Information System (LIS) relative soil moisture products from NASA's Short-term Prediction Research and Transition (SPoRT) Center uses the Noah land surface model, real-time Suomi NPP (National Polar-orbiting Partnership) green vegetation fraction, and radar- and gauge-derived precipitation estimates to generate daily modeled analyses of soil moisture at 0–10 cm and 0–2 m depth. Values of zero percent indicate no moisture in the soil, and values of 100 percent indicate complete saturation. The near-surface 0–10 cm layer responds quickly to heavy rainfall, while the deeper 0–2 m layer represents longer-term water storage.

The Evaporative Stress Index (ESI) is a four-week composite product updated weekly and reveals regions of drought where vegetation is stressed due to lack of water. ESI observes reduced rates of water loss through the use of land surface temperature (LST) before it can be observed through decreases in vegetation health, or "greenness." When the lack of water stresses plants, they reduce their transpiration to conserve water by closing their stomata, leading to elevated leaf temperatures that can be observed from space. Healthy green vegetation with access to plenty of water generally warms at a much slower rate than dry and stressed vegetation. Based on observations of variation changes in land surface temperature, the ESI indicates how the current rate of evapotranspiration compares to normal conditions.

Prefire analysis

The 0–2 m relative soil moisture product on October 8, 2018, shows little moisture across most of California, which remained dry for weeks before the Camp Fire on November 8, 2018. The shallow 0–10 cm relative soil moisture product shows some soil moisture that progressively dried out by October 22. The ESI shows mixed areas of low-to-high evaporative stress throughout October, but after weeks of dry soil weather, the vegetation quickly became stressed across the state, as shown on November 11. The combination of so much dry vegetation, warm temperatures, and dry winds contributed to the quick spread of the Camp Fire. Authorities and decision-makers could use these kinds of soil moisture and evapotranspiration datasets in combination with other forecasts and models. This information would provide a more comprehensive picture to identify areas of concern and determine where to carry out mitigative actions and preparedness efforts.

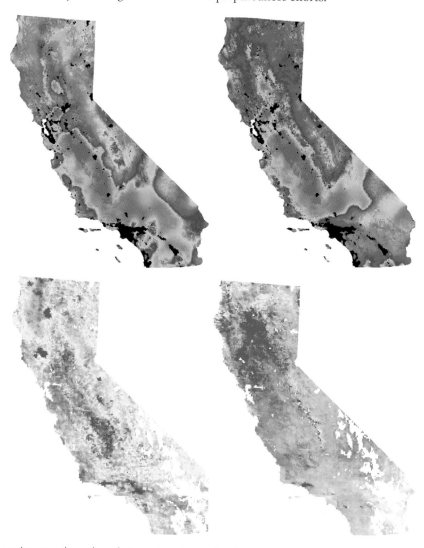

These images show the relative soil moisture for the top 10 centimeters of soil from the Land Information System (LIS) on October 8 and 22, 2018, and the Evaporative Stress Index (ESI) for the weeks ending on October 8 and November 11, 2018. As the top level of soil moisture dried out evapotranspiration decreased, which could have led to more stressed vegetation and an increase in available fuel load.[4-8]

This image of relative soil moisture from the Land Information System (LIS) on October 8, 2018, shows the top 2 meters of soil were very dry weeks before the Camp Fire, which could have led to more stressed vegetation and an increase in available fuel load.[5-8]

During-event and postfire analysis

As the Camp Fire spread across Butte County, California, in November 2018, the ARIA team at NASA's Jet Propulsion Laboratory used SAR to pierce through the dense smoke and detect areas that were likely damaged by the fire. A Damage Proxy Map (DPM) is derived from SAR images from the two Copernicus Sentinel-1 satellites, operated by the European Space Agency. The color variation from yellow to red indicates increasingly significant ground surface change. Copernicus Sentinel-1 uses a C-band SAR, which cannot penetrate dense tree canopies, so underlying structural damage may not be detected in more heavily forested areas.

The California Department of Forestry and Fire Protection (CAL FIRE) provided ground-truthing data during the event, as shown by the colored houses in the figure, allowing NASA to calibrate observations used to create a more accurate DPM. As the fire continued to burn, additional DPMs helped CAL FIRE identify areas that required reinspection. The DPMs are designed to increase situational awareness of potential damage over large spatial areas. The ability to "see" through smoke, clouds, and at night increases the amount of potential information sources during and postevent. Responders can quickly identify critical infrastructure and assets that may have been damaged using the Intersect tool with a vector version of the DPM. These maps are a form of change detection, measuring differences in the pre- and postevent SAR images that represent ground surface disturbance.

The NASA Damage Proxy Maps helped CAL FIRE validate its damage inspection data and isolate areas that needed reinspection. The layers were shared and added to maps in ArcGIS Online and compared to a field-based damage inspection layer for discrepancies. CAL FIRE plans to continue sharing data with NASA during large-scale disasters using the same ArcGIS Online framework in the future.

This Damage Proxy Map shows areas where property was likely destroyed by the Camp Fire. Color variation from yellow to red indicates increasingly more significant ground surface change, which may not be detected under dense tree canopies. The field-collected damage inspection layer from CAL FIRE depicts individual structures as houses, with yellow and red indicating partial or total damage; green indicates undamaged homes, with precious few showing in this view over Paradise, California.[3]

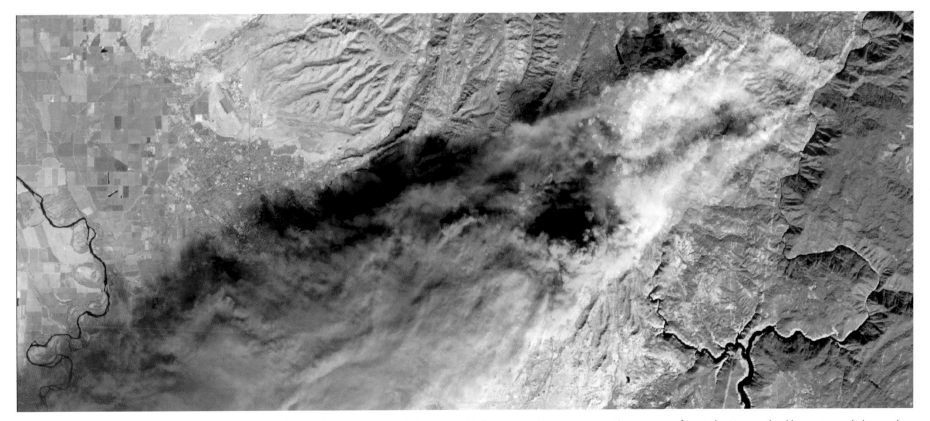

This false color Landsat 8 image of the Camp Fire on November 8, 2018, used the thermal infrared band to show areas that were on fire at the time or had been recently burned. Dense smoke obscures the optical sensor-based image, so SAR was needed during the fire to evaluate conditions on the ground.

2019 MIDWEST FLOODING

Starting in early 2019, much of the US Midwest across the Missouri River and Mississippi River basins experienced widespread flooding that continued through the summer and into the fall in some locations. Many areas saw record or near-record high-water marks, with some gauges above flood stage for more than four consecutive months. Because of the long duration of the flooding and the large geographic area affected, satellite data proved to be invaluable in monitoring and documenting the extent of flooding throughout the event.

Scientists at NASA Marshall Space Flight Center (MSFC) and Goddard Space Flight Center (GSFC) produced water extents and flood maps using optical sensors (MODIS, Sentinel-2, Landsat 8) and SAR (Sentinel-1) to create a composite image of flooding from March through June 2019. Using a collection of different sensors with different spatial and temporal resolutions provided a more complete view of the flooding on days when clouds obscured flooding and when the higher-resolution sensors did not pass over the affected areas.

As the flooding continued through the spring, NASA scientists produced water extent maps that were given to partner organizations such as the Federal Emergency Management Agency (FEMA), the National Guard Bureau, and the US Department of Agriculture (USDA). These updated water extents allowed response organizations to understand where and how the flooding evolved, what roadways and infrastructure were inundated, and where they needed to send new aid.

One of NASA's main partners was the USDA National Agricultural Statistics Service (NASS), which used the water extent raster data to help produce its own products addressing the impact of the flooding on croplands in the region. With flooding inundating so much cropland through spring, these water extents helped organizations make decisions related to food security and crop management.

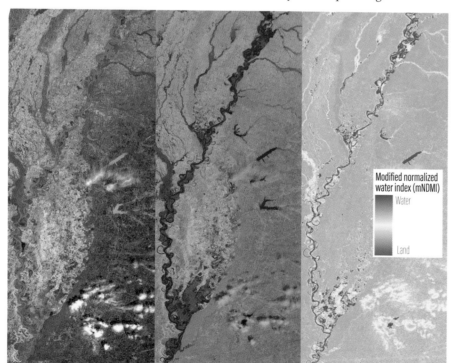

Flooding on the Mississippi River on March 22, 2019, captured using three different views from Copernicus Sentinel-2. Left: True color. Center: Natural color. Right: Modified Normalized Difference Water Index (MNDWI).

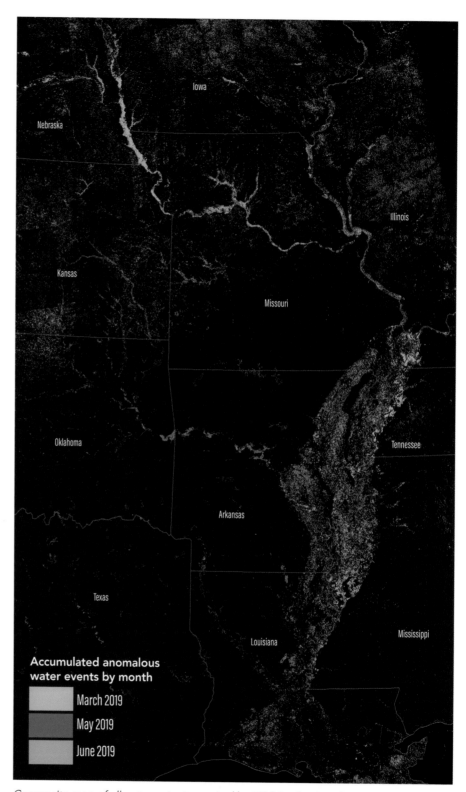

Composite map of all water extents created by NASA scientists showing anomalous water based on the month the images were captured. Composite map used data from Copernicus Sentinel-1 and Sentinel-2, Landsat-8, and MODIS. Images are from March, May, and June 2019. Note: This map is not a complete view of all flooding that occurred over the time period. Some areas may not have been mapped or may have been obscured by cloud cover when satellites passed overhead.

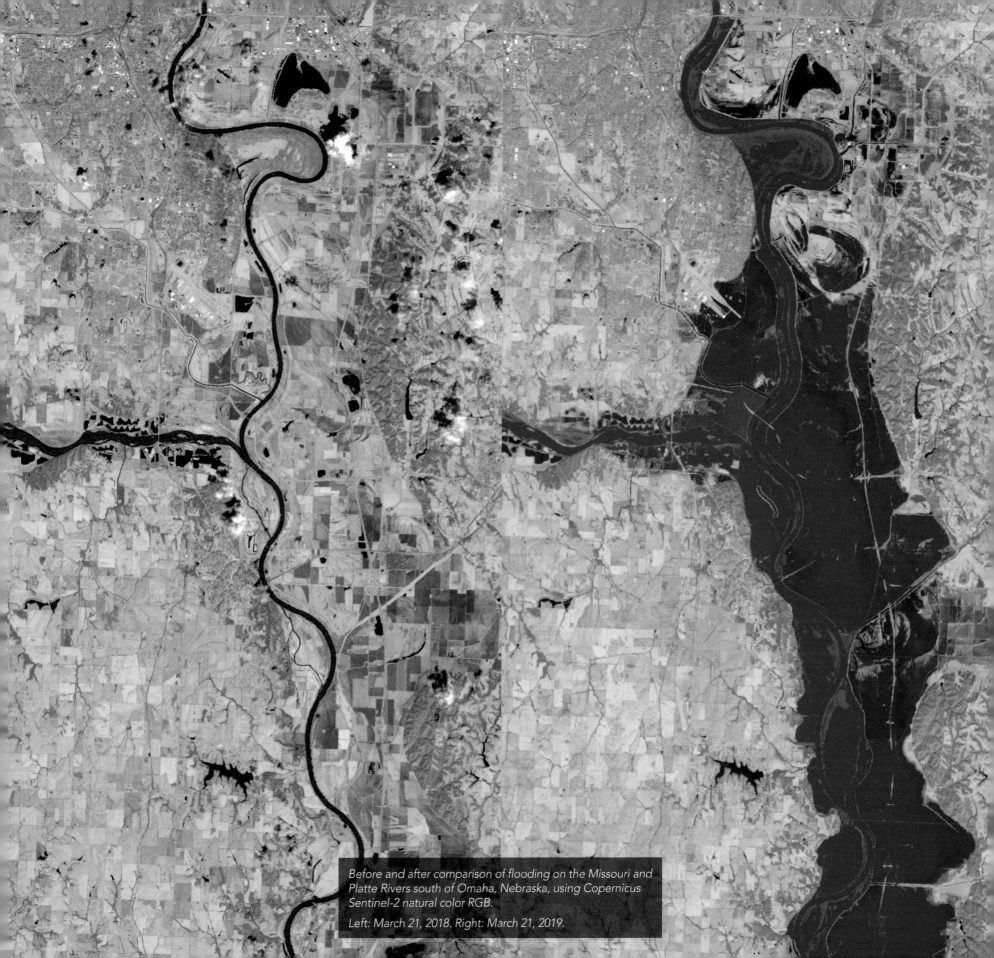

Before and after comparison of flooding on the Missouri and Platte Rivers south of Omaha, Nebraska, using Copernicus Sentinel-2 natural color RGB.

Left: March 21, 2018. Right: March 21, 2019.

HURRICANE MARIA: BLACK MARBLE HD

Hurricane Maria made landfall on Puerto Rico on September 20, 2017, as a Category 4 hurricane, causing widespread destruction across the island, including its infrastructure. Because of the island's terrain and the magnitude of damage to the electrical grid, much of the island suffered from power outages lasting more than six months in some areas, as seen in the large map shown here. Prolonged power outages can result in disruptions of vital services such as medical treatments and procedures, access to medical records, communication, and the storage of goods that require refrigeration.

Black Marble HD, developed by Dr. Miguel Román and a team of NASA scientists from Goddard Space Flight Center (GSFC) and Marshall Space Flight Center (MSFC), uses Visible Infrared Imaging Radiometer Suite (VIIRS) Day/Night Band (DNB). These data were processed and corrected to filter out stray light from the moon, fires, airglow, and any other sources that are not electric lights. Their processing techniques also remove as much other atmospheric interference—such as dust, haze, and thin clouds—as possible. To make the VIIRS data more useful to first responders, the team scaled the observations onto a basemap and incorporated high-resolution GIS data from OpenStreetMap to emphasize locations of streets and neighborhoods.

Ongoing research and development of the Black Marble HD continues under the guidance of Dr. Román, director of the Earth from Space Institute at the Universities Space Research Association, where the product continues to show value in other recent disaster response and recovery scenarios.

Understanding what areas of Puerto Rico were still experiencing power outages can help decision-makers monitor the long-term recovery efforts across Puerto Rico. This understanding helps responders to identify locations that need additional resources, have compromised logistical infrastructure leading to lack of fuel and supplies, are more at risk from cascading effects such as food insecurity, and are vulnerable to impacts from compounding hazards, such as earthquakes or additional hurricanes.

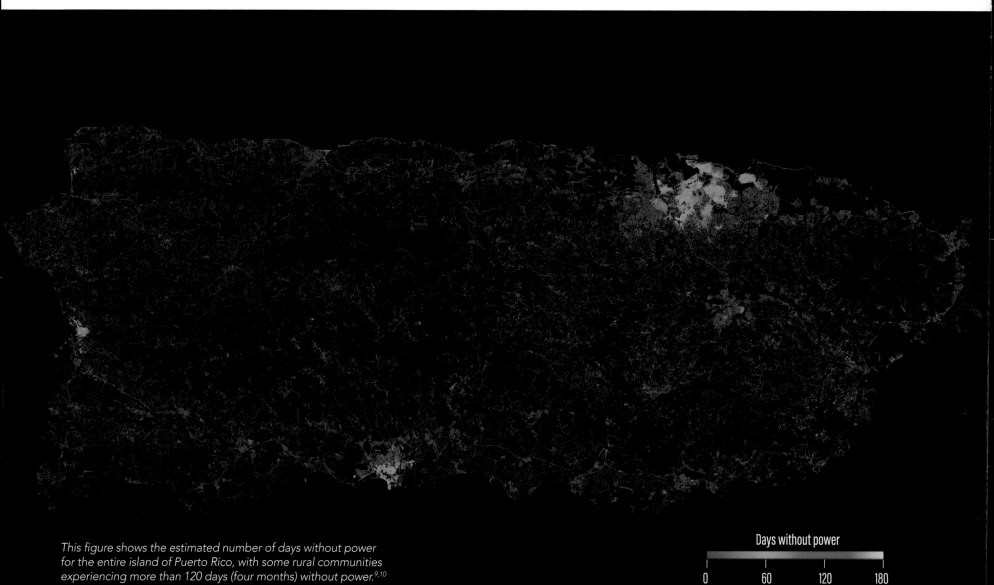

This figure shows the estimated number of days without power for the entire island of Puerto Rico, with some rural communities experiencing more than 120 days (four months) without power.[9,10]

Days without power

0   60   120   180

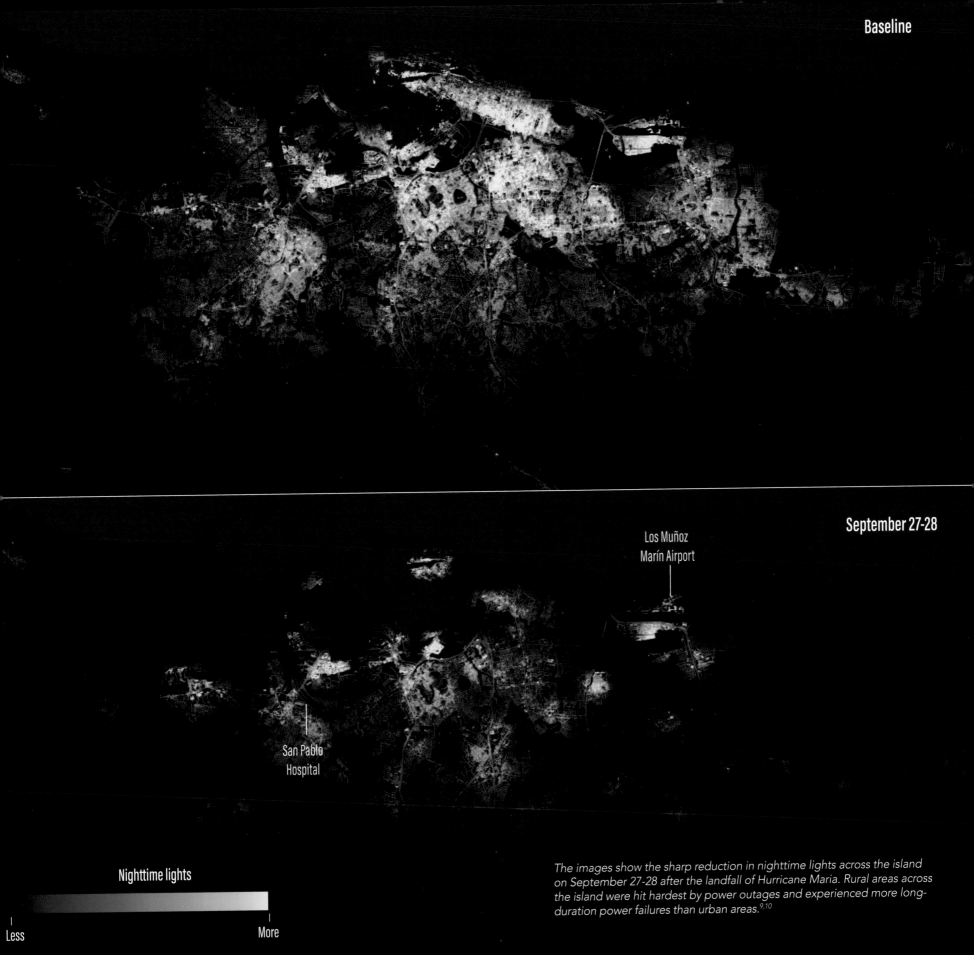

Baseline

September 27-28

Los Muñoz
Marín Airport

San Pablo
Hospital

Nighttime lights

Less

More

The images show the sharp reduction in nighttime lights across the island on September 27-28 after the landfall of Hurricane Maria. Rural areas across the island were hit hardest by power outages and experienced more long-duration power failures than urban areas.[9,10]

AUSTRALIA FIRES: SMOKE PLUME DATA

On December 16, 2019, NASA's Terra satellite flew over the eastern coast of Australia, capturing the height of smoke plumes emanating from the fires with its Multi-angle Imaging Spectroradiometer (MISR) instrument. The original 2D raster showed heights up to 4.5 km above Earth's surface, but the structure of the plumes was difficult to visualize.

For the first time, MISR data were converted to 3D and visualized in an interactive web application by extracting point values from the 2D raster into a comma-separated values (CSV) file and plotting the point data in 3D. This conversion allowed users to view the plumes from many angles to see how they varied in structure.

These Earth-observing satellite data provided researchers and disaster management agencies with the "big picture" of the location and intensity of fires in the region and gave an idea of where the smoke was being transported. The data can also be used to initialize air-quality and chemical transport models. In particular, plume elevation data from MISR can greatly improve the accuracy of models in predicting where the smoke will go and what regions may be affected downwind.

While the data were visualized after the event, the Australian Bureau of Meteorology (BOM) is incorporating MISR data and other NASA data sources into its air-quality models to improve accuracy for future forecasts.

The plume heights in this image are represented as spheres, with progressively lighter colors for higher elevation. The height has been visually exaggerated 20 times to better see the details in the data. The 2D raster is shown on the ground for comparison. In addition, hot spot data from the Terra satellite's MODIS instrument are shown as red spheres on the ground, indicating areas of active fires. The base layer of natural-color imagery is from MISR's nadir-viewing camera.

NOTES

1. Stanley, T., and D. B. Kirschbaum (2017), "A Heuristic Approach to Global Landslide Susceptibility Mapping." *Natural Hazards*, 1–20, doi:10.1007/s11069-017-2757-y.

2. Kirschbaum, D., and Stanley, T. (2018), "Satellite-Based Assessment of Rainfall-Triggered Landslide Hazard for Situational Awareness. Earth's Future." doi:10.1002/2017EF000715.

3. Yun, S., Hudnut, K., Owen, S., Webb, F., Simons, M., Sacco, P., Gurrola, E., Manipon, G., Liang, C., Fielding E., Milillo, P., Hua, H., Coletta, A. (2015), "Rapid Damage Mapping for the 2015 M7.8 Gorkha Earthquake Using Synthetic Aperture Radar Data from COSMO-SkyMed and ALOS-2 Satellites." *Seismological Research Letters*, Vol. 86(6), 1549-1556, doi: 10.1785/0220150152.

4. Anderson, M. C., C. R. Hain, B. Wardlow, J. R. Mecikalski, and W. P. Kustas (2011), "Evaluation of a Drought Index Based on Thermal Remote Sensing of Evapotranspiration over the Continental US." *Journal of Climate*, 24, 2025–2044.

5. Blankenship, C. B., J. L. Case, W. L. Crosson, and B. T. Zavodsky (2018), "Correction of Forcing-Related Spatial Artifacts in a Land Surface Model by Satellite Soil Moisture Data Assimilation." *IEEE Transactions on Geoscience and Remote Sensing*,15(4), 498–502. doi: 10.1109/LGRS.2018.2805259.

6. Case, J. L. (2016), "From Drought to Flooding in Less than a Week over South Carolina." *Results in Physics*, 6, 1183–1184.

7. Case, J. L., Zavodsky, B. T., (2018), "Evolution of 2016 Drought in the Southeastern United States from a Land Surface Modeling Perspective." *Results in Physics*, 8, 654–656.

8. Case, J. L., White, K. D., Guyer, B., Meyer, J., Srikishen, J., Blankenship, C. B., Zavodsky, B. T., (2016). "Real-Time Land Information System over the Continental US for Situational Awareness and Local Numerical Weather Prediction Applications." 30th Conf. Hydrology, New Orleans, LA., Amer. Meteor. Soc., 3.3.

9. Román, M.O. et al. (2019), "Satellite-Based Assessment of Electricity Restoration Efforts in Puerto Rico after Hurricane Maria." *PLoS One*, 14 (6).

10. Román, M.O. et al. (2018), "NASA's Black Marble Nighttime Lights Product Suite." *Remote Sensing of Environment* 210, 113–143.

Image credits

Pages 150–151—Hurricane Florence: NASA Earth Applied Sciences Disasters Program (IMERG); NASA Precipitation Measurement Missions (PMM) Science Team, The Global Precipitation Measurement (GPM) Mission, NASA, Japan Aerospace Exploration Agency (JAXA).

Pages 154–155—Huricane Willa: NASA Earth Science

Pages 156–157—Rainfall Accumulation: NASA Precipitation Measurement Missions (PMM) Science Team, The Global Precipitation Measurement (GPM) Mission, NASA, Japan Aerospace Exploration Agency (JAXA). Text contribution by Jim Schultz.

Page 158—Spyglass Application: Sentinel-1 data were accessed through the Copernicus Open Access Hub. The image contains modified Copernicus Sentinel data (2018), processed by ESA and analyzed by the NASA-JPL/Caltech ARIA team. Flood Proxy Map: Sentinel-1 data were accessed through the Copernicus Open Access Hub. The image contains modified Copernicus Sentinel data (2018), processed by ESA and analyzed by the NASA-JPL/Caltech ARIA team.

Pages 159–161—Landsat-8 Camp Fire Image and Northern California Camp Fire: NASA Earth Observatory image by Joshua Stevens, using Landsat data from the US Geological Survey, and MODIS data from NASA EOSDIS/LANCE and GIBS/Worldview.

Page 161—ARIA Damage Proxy Map (DPM): Sentinel-1 data were accessed through the Copernicus Open Access Hub. The image contains modified Copernicus Sentinel data (2018), processed by ESA and analyzed by the NASA-JPL/Caltech ARIA team. This research was carried out at JPL funded by NASA.

Page 162—MODIS Flood Map: NASA GSFC Flood Mapping Project.

Pages 162–163—Midwest Flooding 2019: The Sentinel data used in this derived product contains modified Copernicus Sentinel data (2019), processed by ESA, Alaska Satellite Facility, the US Geological Survey (USGS) and NASA Marshall Space Flight Center. Landsat-8 imagery courtesy of the US Geological Survey (USGS) and NASA Marshall Space Flight Center.

Pages 164–165—Hurricane Maria: Image and Text Credits, NASA Black Marble Team, NASA Science Visualization Studio, NASA Earth Observatory. Sentinel data used in this derived product, contains modified Copernicus Sentinel data (2019), processed by ESA, Alaska Satellite Facility, and NASA Marshall Space Flight Center.

Page 166—Australia Fires: MISR data from the Active Aerosol Plume-height (AAP) project, V. Flower and R. Kahn (NASA GSFC); 3D rendering by J. Kirkendall (NASA HQ); Text by Jacob Reed (NASA GSFC).

Acknowledgments

The authors would like to thank the many people who contributed to this chapter with images, edits, and feedback: David Green, Jordan Bell, Kenton Fisher, Verity Flower, Michael Goodman, Christopher Hain, Brady Helms, Ralph Kahn, Aries Keck, Dalia Kirschbaum, Ronan Lucey, Andrew Molthan, John Murray, Frederick Policelli, Jacob Reed, Amy Robinson, Miguel Román, Lori Schultz, Dan Slayback, Thomas Stanley, Will Stefanov, Sang-Ho Yun, Dartmouth Flood Observatory, CAL FIRE.

ENVIRONMENTAL MONITORING WITH DRONES AND GeoAI

The blue catfish is an invasive species that is wreaking havoc in Chesapeake Bay. Scientists are using drone imagery, artificial intelligence (AI), and GIS as they probe to understand the full scope of the problem.

By William Shuart, **Virginia Commonwealth University;** and Rohit Singh, Lain Graham, and Gerald Kinn, **Esri**

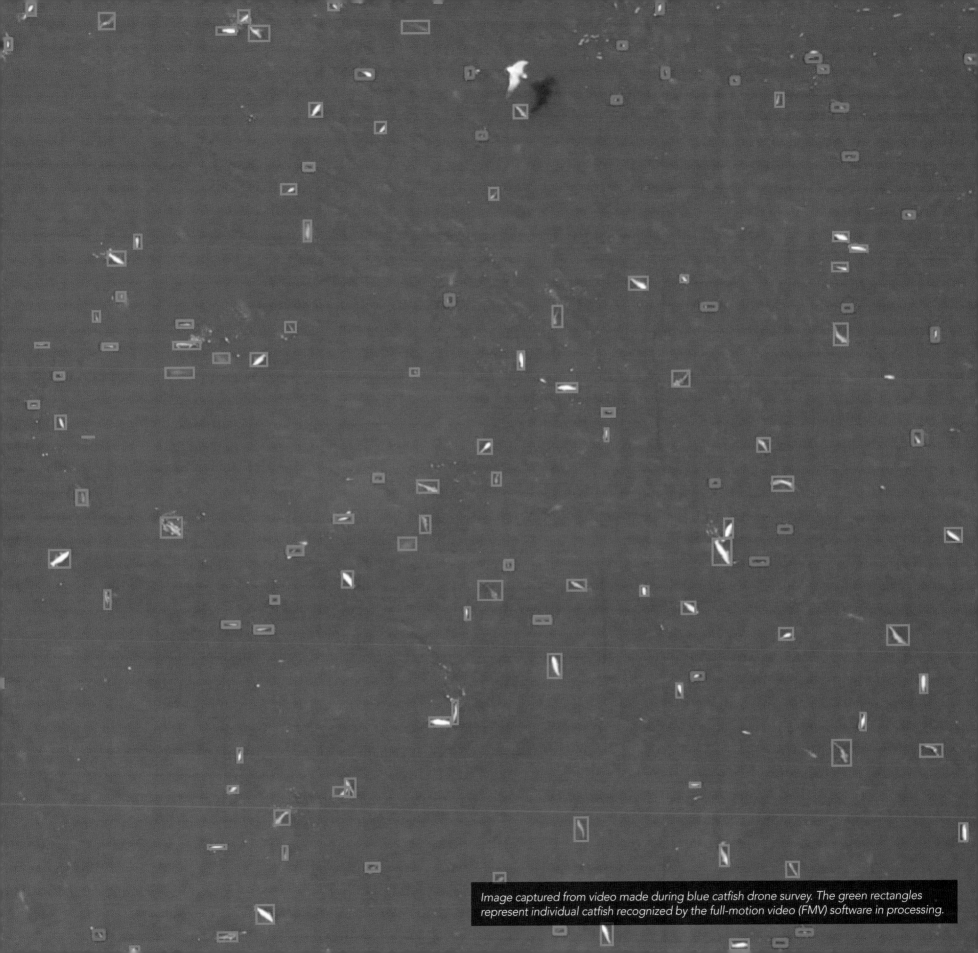

Image captured from video made during blue catfish drone survey. The green rectangles represent individual catfish recognized by the full-motion video (FMV) software in processing.

WHY AI AND DRONES?

The application of small uncrewed aerial vehicles (UAVS; also known as drones) to geospatial investigations has exploded in just a few years' time. The new ability to place a camera several hundred feet above the surface is much less costly than using traditional aircraft and delivers data in extraordinarily high resolution. Drones are highly maneuverable and can carry a wide variety of imaging sensors. Thus, they return accurate and actionable information and can solve time-sensitive problems that require high resolution.

This is a story about blue catfish and how to count them using drones. A notorious invader to the Chesapeake Bay region, the species has had an outsized impact on that watershed's ecosystem. In the context of this chapter, we will go "back to the future," introducing geospatial data collected via drones. These extremely large datasets are big data, and distilling them requires machine learning, artificial intelligence (AI), and GIS software.

In this example, instead of attempting to collect and count fish from a boat in real time or even take low-angle photos to count from later, researchers achieved a literal bird's-eye view using drones. Specifically, drones view and capture data that researchers can process and review back in a lab for accuracy. Gathering these data has become a simple and repeatable process, but the sheer volume of data presents a new issue. But the ease in gathering so much data so often also requires new methods for reviewing it all. Fortunately, automated intelligence—specifically the fusion of GIS, AI, and GeoAI—unlocks these massive datasets for review, resulting in deeper understanding.

The term *GeoAI* is used to describe the use of artificial (and automated) intelligence to solve a geospatial problem. This chapter discusses the problem of catfish as an invasive species outside of their native watershed. To address this geospatial problem, researchers examined the size of a catfish population in place and time.

This emerging application of drone-based imagery has a new name: *computer vision*. It uses images and video to train computers to see what the human eye can see, with the goal of achieving higher-level understanding.

The blue catfish (Ictalurus furcatus) *is the largest species of North American catfish. The fish is considered an invasive pest in some areas, particularly the Chesapeake Bay. Because blue catfish tolerate brackish waters, it can colonize inland waterways in coastal regions.*

INTRODUCED SPECIES

Species migration and colonization of new habitats through natural dispersion or movement processes have allowed species to expand their range and colonize new ecosystems for millions of years. The ability for plant seeds to be dispersed, fish to tolerate flooding and salinity changes, and birds to fly thousands of miles enables species to limit competition for resources and increase populations. However, during the past 200 years, humans have accelerated this process—sometimes purposely transporting species from one location to another. Examples include introducing an animal for harvesting (nutria), birds and frogs for controlling bugs (European starling and cane toad), plants for adorning gardens (kudzu), and fish for sportfishing and food (blue catfish).

When a species is moved or is found outside of its normal range, it is referred to as introduced. Species that are introduced into a new location or environment may have the ability to outcompete native species or possess abilities to change or modify that new habitat or environment. This is referred to as an invasive species. Species introductions sometimes can be unintentional, or passive, such as when ships take on ballast water for buoyancy in one location, travel to another location, and then deposit that water (and everything in it) with potentially new and invasive organisms into a new location and environment. This process and mechanism are how zebra mussels (Dreissena polymorpha) were introduced into the Great Lakes and have dramatically changed the Great Lakes ecosystem in many negative ways. The introduction of nonnative (and usually invasive) species is sometimes referred to as biological pollution. The US map of nonnative species highlights the extent of that distribution showing the number of occurrences in each county of introduced species.

However, humans have also performed strategic introductions of species into new environments, for example, to control other species—some that are even introduced themselves. For example, grass carp (Ctenopharyngodon idella) were introduced in the 1960s in New Zealand to control invasive aquatic vegetation and have since been used in 45 of the 50 US states. We also introduce species for food sources through aquaculture processes and for recreational purposes.

Definite and unsurprising patterns emerge when invasive species are mapped across the contiguous United States.

INVASIVE CATFISH

Blue catfish are native to the Mississippi, Missouri, and Ohio River basins of the central and southern United States, where they support both recreational and commercial fisheries (see catfish range map). Beginning in 1974, the Virginia Department of Game and Inland Fisheries introduced more than 300,000 juvenile blue catfish into coastal rivers of Virginia to establish self-sustaining recreational and commercial fisheries, starting with the James and Rappahannock rivers, and ending in 1985 with introductions into the York River system. Because blue catfish can live longer than 20 years, weigh more than 100 pounds, grow longer than 4 feet, and have a unique taste, they have been introduced into many rivers, lakes, and estuaries throughout the United States.

Blue catfish have a vast salinity tolerance of almost half seawater (15 parts per trillions for 72 hours), enabling them to survive and reproduce in freshwater riverine and estuarine systems. They are opportunistic predators, and familiar prey include macroinvertebrates, blue crabs, and many other fish species. Studies have shown that blue catfish have taken the apex predator spot, feeding higher on the food chain than striped bass and other predators in the Chesapeake Bay since their introduction.

Blue catfish also support important recreational fisheries, including a nationally recognized trophy fishery in the James River, Virginia, where one-third of total recreational fishing effort for freshwater species is directed at catfish. The species represents a large support base for recreational fishing and income for the agencies managing the fishery stocks. Blue catfish mature at about 2-feet long or about three years of age. They have a high reproduction rate and continue to grow in length and weight as long as they live.

Recreational fishing license sales increased in Virginia, and commercial harvests also increased dramatically since the 1970s. However, the introduction of the species significantly altered the ecosystems of the estuaries and lower Chesapeake Bay rivers.

All the reasons why blue catfish were originally stocked in Virginia are also the reasons why the species has become invasive in the Chesapeake Bay ecosystem. Since the introduction of blue (and flathead) catfish, an additional trophic layer was added, because these introduced predators feed off several groups.

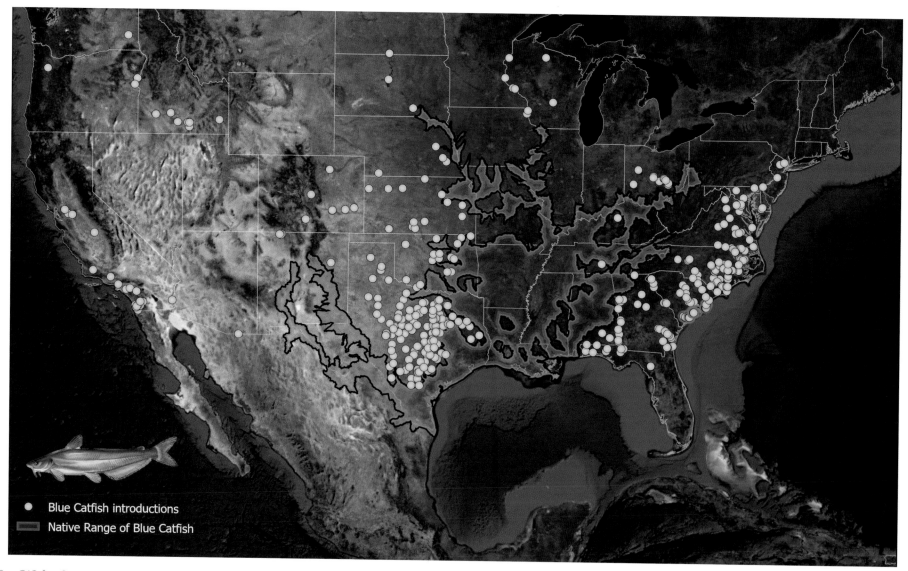

● Blue Catfish introductions

▭ Native Range of Blue Catfish

THREATENING NATIVE SPECIES

The introduction of blue catfish into Chesapeake Bay brought thousands of fish native to the Mississippi River into the Rappahannock and James Rivers. But the nutrient-rich habitat was *too* good for the blue catfish. They became abundant and navigated into waters where they had never been recorded before.

The term *trophic position* refers to the position an organism occupies in the food web. Blue catfish eat almost anything—including blue crabs, insects, plants—that native fish need to survive, and they also eat other fish. For these and other reasons, their relatively high trophic position in the Chesapeake Bay watershed threatens the ecology and economy of the highly valued waters. The Maryland Biological Stream Survey and the Virginia Healthy Waters Program have identified and mapped these waters and the threats to the ecosystem brought on by the explosion of blue catfish populations.

The intentional introduction of the blue catfish for sport fishing in the 1970s (opposite page) has degraded the Chesapeake Bay marine ecosystem by greatly reducing the presence of many native species.

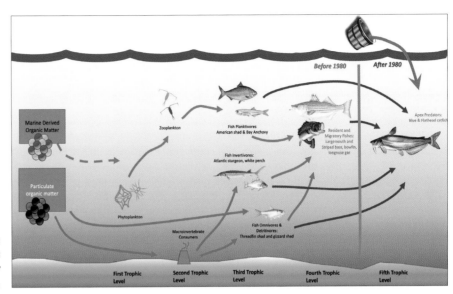

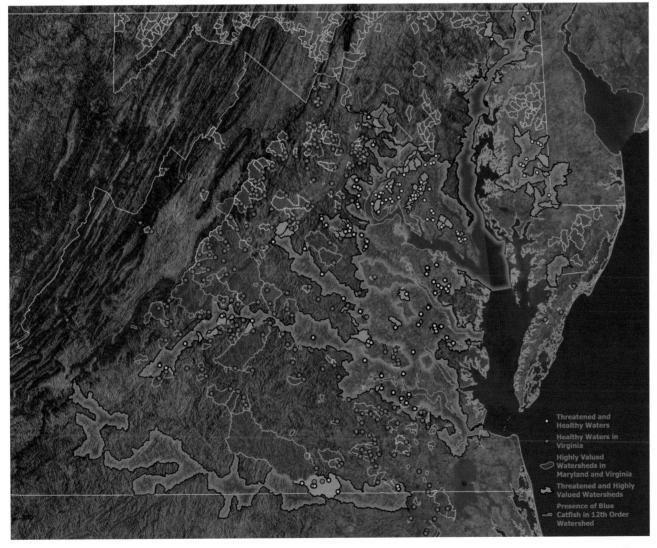

Invasive species threaten healthy ecosystems. This map shows the high-valued aquatic ecosystems in Maryland and Virginia and the current extent of blue catfish.

TRADITIONAL DATA COLLECTION

Catfish are demersal—they live near the bottom—and collecting data on catfish involves several kinds of equipment and methods. Trawling involves dragging a net behind a boat at a certain depth, bringing the net up, and then sorting the different species and counting the individual catfish. While this method works, you can't target just one or two species of interest. You must bring up all the fish, including potentially threatened and rare species.

This method stresses all fish and takes time to sort out the species. A newer method is called *low-frequency electrofishing*, which, despite the name, is harmless to fish. With specialized equipment and a special power generator on board, a boat crew can place low-frequency and low-voltage current in the water. Because catfish live on the bottom, they respond to the electric current by swimming off the bottom to the surface. Scientists then use nets to pull fish onto the boat. Electrofishing allows scientists to target only catfish rather than other types of fish species, limiting interactions with other potentially threatened and endangered species.

Commercial and recreational fisheries

A current effort in Virgina allows low-frequency electrofishing to harvest blue catfish. This effort has proved effective and profitable while not harming other species. Commercial landings of blue catfish throughout the past 30 years have increased.

Because blue catfish are now the apex predators in the Chesapeake Bay estuary, they are also susceptible to environmental contamination and biomagnification. Several harmful environmental disasters have occurred in the James River that released polychlorinated biphenyl (PCBs), Kepone (an insecticide), and other chemicals that biomagnify (increase in concentration) as the trophic level increases. Blue catfish have shown increased levels of these and other contaminants. The Virginia Department of Health has set a limit on how much blue catfish an individual can consume: fewer than two servings per month, and in some areas they recommend not eating them at all. However, many recreational fishers consume blue catfish as part of their sustenance. "Catfish Fry" gatherings are an enduring regional tradition.

Population size, age, and weight

After collecting the fish, researchers record the length and weight of each fish and create "growth curves" for individual species. If researchers collect enough data, they can estimate the health of specific fish populations. Each species has a unique curve, which can change depending on location, diet, and stress. For example, blue catfish can weigh upward of 90 pounds (the Virginia record is 143.3 pounds), but they are not very long fish. Fish weighing 50 to 80 pounds are not uncommon.

To determine the age of a fish, researchers can also count *annulus* rings on either the spine or the inner ear bone (called an *otolith*). This method uses the same principles as counting the rings to age trees. The data collection is labor-intensive. Every fish must be examined. Historically, this method has been the only way to collect the data required to make informed decisions about fish stocks. Compounding the collection problem, researchers cannot achieve accurate results because it is difficult to catch, measure, and weigh a certain individual fish species in an area of interest. Electrofishing causes catfish to rise to the surface over a wide area, but the boat can cover just a small portion of that area.

Enter the drone.

Blue catfish grow rapidly. The specimen on the right is approximately 12 months old.

Low-frequency emitter

Electrofishing research vessel in operation.

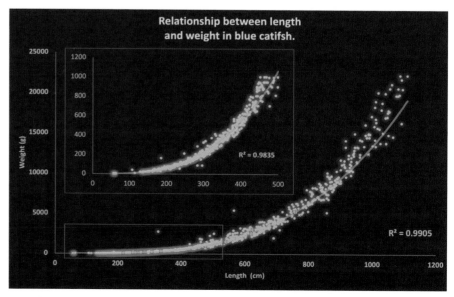

Growth curve of the blue catfish.

INTEGRATING DRONES IN FIELD SCIENCE

A drone flying at low altitude can capture images of surfacing catfish with relative ease, technically speaking. The challenge is to convert that raw imagery into quantifiable data.

For this groundbreaking effort, a proof of concept used two different UAV platforms—the 3DR Solo equipped with a GoPro Hero camera and a DJI Mavic Pro 2 with a built-in camera—to obtain videos that can be multiplexed with the plug-in ArcGIS Full Motion Video. Multiplexing in this context is the process of combining the two files containing the video and metadata files to, in effect, *georeference* the video frames.

Videos of commercial electrofishing and fish collection were formatted and then multiplexed to insert the time stamp, location, and orientation into the videos. Researchers can then read, play, and digitize the multiplexed movies in ArcGIS Pro. Researchers can count, measure, and do other kinds of things frame by frame and save those edits to a geodatabase. But thousands of fish are seen every second in the video, so they had to come up with a different way to analyze them: filtering. To borrow from Esri's latest campaign, *See What Others Can't* also implies only seeing what you want or need from a dataset. In this case, researchers were interested in catfish, not birds or waves or the boat.

But even the process of counting and measuring each fish manually on a computer took too much time to be efficient. Fortunately, AI and machine learning have made rapid progress in recent years. Computer vision—the ability for computers to see—is now real, cost effective, and viable. Researchers can use deep convolutional neural networks (deep learning) to automate the task of detecting catfish from the drone videos and provide estimates of their size, and hence their age and growth statistics. In computer vision, this process is known as *object detection*. Deep-learning-based,

object-detection models can detect objects of interest in imagery and report their location in terms of bounding boxes. Researchers can use bounding boxes to estimate the size of each fish, and the number of boxes represents their population. Since the video is geospatially enabled with the flight log, sensor dynamics, and field of view (FOV) information, researchers can translate these measurements from image space to map space, thereby enabling analysis within the correct geographical context.

Deep learning "learns" by looking at multiple examples of objects that it needs to recognize. The team used ArcGIS Pro to mark the location and size of each catfish in several frames of the georeferenced video. This process served to train the deep learning model.

However, these data cannot be directly fed into the deep learning model. Training deep convolutional neural networks with millions of parameters is computationally expensive and is typically performed on GPUs that have limited memory capacity. So the training data are fed into the models and processed on the GPUs in small batches, consisting of subimages, also known as image chips, along with their labels, that is, the attribute about the objects contained within those chips and their bounding box locations.

The team used the Export Training Data for Deep Learning tool in ArcGIS Pro to export training samples in the PASCAL_VOC_rectangles (Pattern Analysis, Statistical Modeling, and Computational Learning, Visual Object Classes) format. This PASCAL VOC dataset is a standardized image dataset for object class recognition. The label files are XML files and contain information about image name, class value, and bounding boxes.

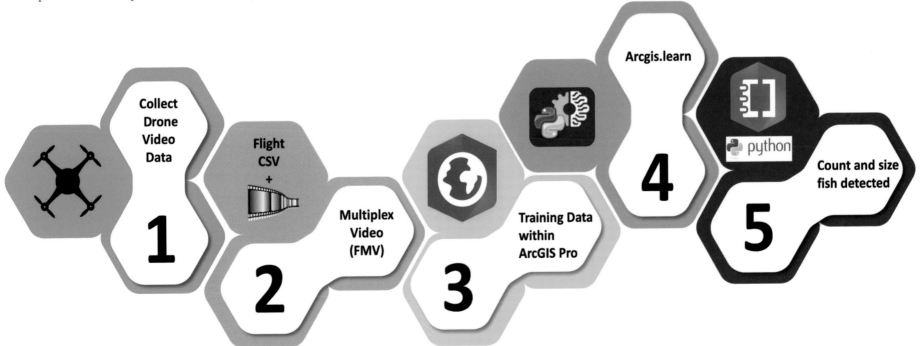

Workflow from drone data collection culminating with count and size detection results.

TRAINING THE CATFISH DETECTOR

In the next step, researchers used Jupyter Notebooks and the arcgis.learn module in the ArcGIS API for Python to train the catfish detection model. The arcgis.learn module (not to be confused with the teaching site ArcGIS Learn) is built on top of fast.ai and PyTorch and enables the training of highly accurate models with a few lines of code. The type of model trained here was the SingleShotDetector, so-called because of its ability to find all objects in an image (chip) in one pass of the convolutional neural network through the image. We customized the model by having it use a ResNet101 backbone, as opposed to a standard ResNet34 backbone. This convolutional neural network is more powerful, consisting of 101 layers that allowed the team to train a more accurate model.

Additional scripting applied data augmentation techniques, such as randomly zooming, rotating, and flipping the images, which enabled the training of a model with limited data and thus better generalizing over unseen images.

Deep learning models must be initialized with a learning rate. Researchers must set the value of this important hyperparameter before the learning process begins. The learning rate determined how the researchers adjusted weights for their network concerning loss gradient. The ArcGIS.learn module uses fast.ai's learning rate finder to find an optimal learning rate for training models. They trained the model over 300 epochs (or passes through the entire training dataset), using the suggested learning rate. The trained model could then detect catfish fairly well, as seen in the side-by-side visualized results.

Three distinct technologies—electrofishing, machine learning, and UAVs—combine to turn chaotic video into quantifiable blue catfish population data.

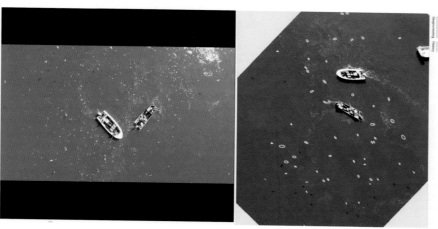

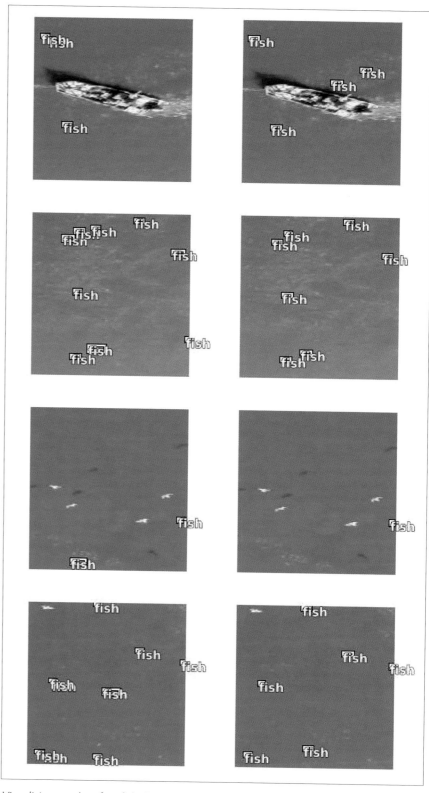

Visualizing results of catfish detector. The ground truth is shown on the left and the model's predictions on the right. In some cases, the model has detected catfish that were inadvertently missed while labeling them.

DETECTING CATFISH

Researchers used the trained model to detect catfish in georeferenced video frames using the Detect Objects Using Deep Learning tool in ArcGIS Pro. Python API additionally can apply the trained model to a video and save the detected features as Video Moving Target Indicator (VMTI) graphics in the multiplexed full-motion video (FMV). This enabled the team to visualize the detected catfish in their geographical context. Additionally, a CSV file containing the location and size of the detected catfish in each frame of the video was created, enabling the performance of downstream analysis tasks, such as estimating the catfish population and inferring their age and growth statistics.

Time and resources

Once the drone video is multiplexed and brought into ArcGIS Pro, users can perform many functions, such as measuring and adding points to a geodatabase. From a video at 30 frames per second and a length of 8 minutes, a researcher would have to examine 14,400 frames to count and measure individual fish. The GeoAI process provides a method that gives results in a few hours compared to several people spending potentially weeks manually counting using the same video. Now, the actual science process can keep up with data acquisition, as opposed to videos sitting in storage and not being maximized.

Going back to the future

The population estimates for fisheries species are only as good as the data that researchers and managers produce. Researchers at Virginia Commonwealth University have studied blue catfish populations and their impacts on the Chesapeake Bay ecosystem for more than 30 years. The studies found declines in native fish species and replacement by other species, bioaccumulation of heavy metals in blue catfish, and other impacts, yet recreational and commercial fishing thrives. The challenge has always been to produce an accurate estimate of the size and age of catfish populations to provide baseline data on which to base management decisions.

Drones now serve as a tool to supplement the current work of accurately detecting, counting, and sizing of catfish. Using a combination of accurate hands-on work with the latest technologies, researchers can use data developed from the GeoAI process to determine the abundance and growth of blue catfish in the ecosystem.

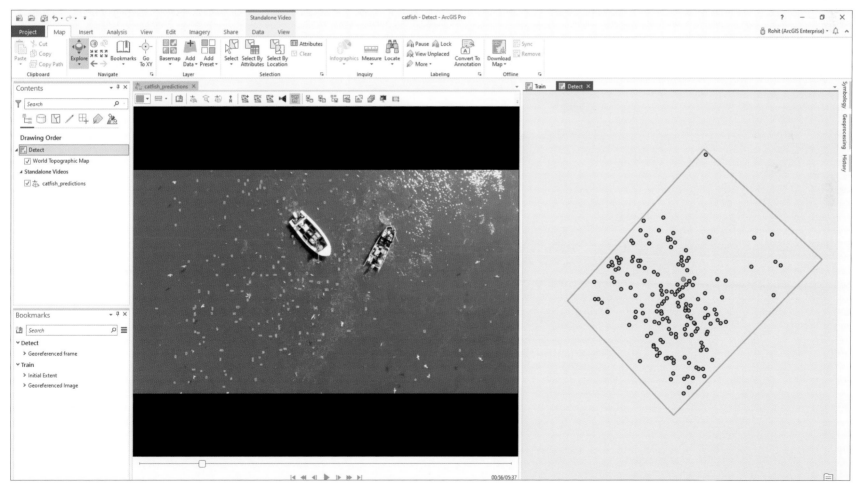

Fish counts extracted from video imagery as seen in ArcGIS Pro interface. The detector picks up most of the fish in each frame of the video (30fps) and detects nearly all fish that come in contact with the surface.

DIRECT DATA — WHY DOES IT MATTER TO STAKEHOLDERS?

Federal and state government agencies

Natural resource (fisheries) biologists, managers, and planners

Information: statistics on fishery effort; catch, harvest, recruitment, and population growth, ecological benefits from invasive species.

Researchers

Academic and agency biologists, ecologists, and environmental managers

Information: basic biology and ecology, invasive species biology, interactions with native species, and restoration potential of invasive species removal by commercial harvest.

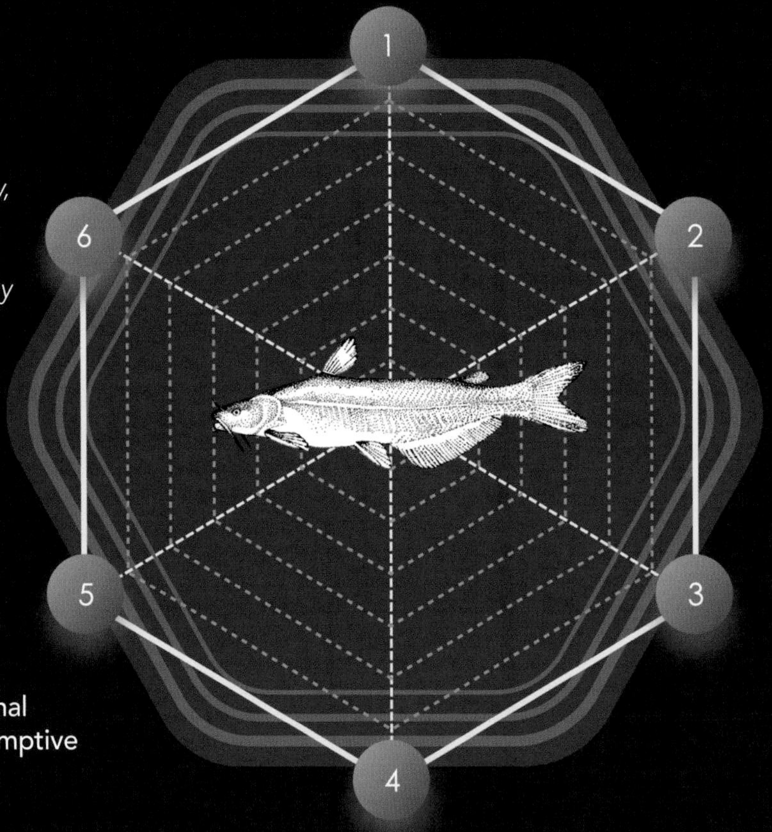

Nongovernmental organizations

Environmental education and awareness, citizen science, and environmental advocacy

Information: effects of invasive predators (e.g., blue and flathead catfish) and fishery policies on Chesapeake Bay living resources.

Commercial fisheries

Commercial fishers, regulating agencies, and related economic interests

Information: blue catfish seasonal and geographic distribution, relative abundance, gear efficiency and bycatch, effect of regulations on profitability and sustainability, and effect of invasive species on commercially important native fisheries.

Recreational fisheries

Fishing guides, licensed recreational fishers, ecotourism and nonconsumptive users, related economic interests

Information: avoiding user conflict with commercial fishers, population growth and size structure, effect of fishing regulations, and effect of invasive species on native fishes.

Seafood industry

Seafood processors, marketers, distributors

Information: potential demand, product quality and sustainability, and marketing as an environmentally friendly product.

NOTES

Garman, G. C., Hale, R., Unger, M., and Rice, G. (1998). *Fish Tissue Analysis for Chlordecone (Kepone) and Other Contaminants in the Tidal James River, Virginia.* A report to the United States Environmental Protection Agency, Washington, DC.

Hilling, C. D., Bunch, A. J., Orth, D. J., and Jiao, Y. (2018). "Natural Mortality and Size-Structure of Introduced Blue Catfish in Virginia Tidal Rivers." *Journal of the Southeastern Association of Fish and Wildlife Agencies*, 5, 30-38.

MacAvoy, S. E., Macko, S. A., McIninch, S. P., and Garman, G. C. (2000). "Marine Nutrient Contributions to Freshwater Apex Predators." *Oecologia*, 122(4), 568-573.

Schloesser, R. W., Fabrizio, M. C., Latour, R. J., Garman, G. C., Greenlee, B., Groves, M., & Gartland, J. (2011). "Ecological Role of Blue Catfish in Chesapeake Bay Communities and Implications for Management." www.gbif.org.

Data sources

University of Georgia—Center for Invasive Species and Ecosystem Health.

Virginia Commonwealth University—INSTAR (https://gis.vcu.edu/instar).

Virginia Healthy Waters Program—https://www.dcr.virginia.gov/natural-heritage/healthywaters.

USGS Nonindigenous Aquatic Species database: USGS Nonindigenous Aquatic Species database Dataset homepage. Citation Neilson M (2019). USGS Nonindigenous Aquatic Species database. Version 1.122.

Find this StoryMap and other resources at www.GISforScience.com.

PART 4
TECHNOLOGY SHOWCASE

This book has already shown how science goes hand in hand with technology (engineering). One of the most exciting trends of the modern age is how science uses the exponential power and assistance of artificial intelligence (AI) to help address the unprecedented challenges facing humanity and the planet, including climate change, water scarcity, global health crises, food security, and loss of biodiversity. GIS technology is no different; it extends our minds by abstracting our world into knowledge objects that we can create, replicate, and maintain. These knowledge objects include data, imagery, and models that explain processes and workflows, as well as maps that communicate and persist in apps. Enjoy this section of vignettes on GIS technologies that help create new systematic frameworks for scientific understanding.

The world according to Spilhaus. Currents derived from maps.com
"Major Ocean Currents" source feature layer ArcGIS Living Atlas of the World.
Basemap from NASA Visible earth bathymetry and natural earth land features.

THE SPILHAUS WORLD OCEAN MAP

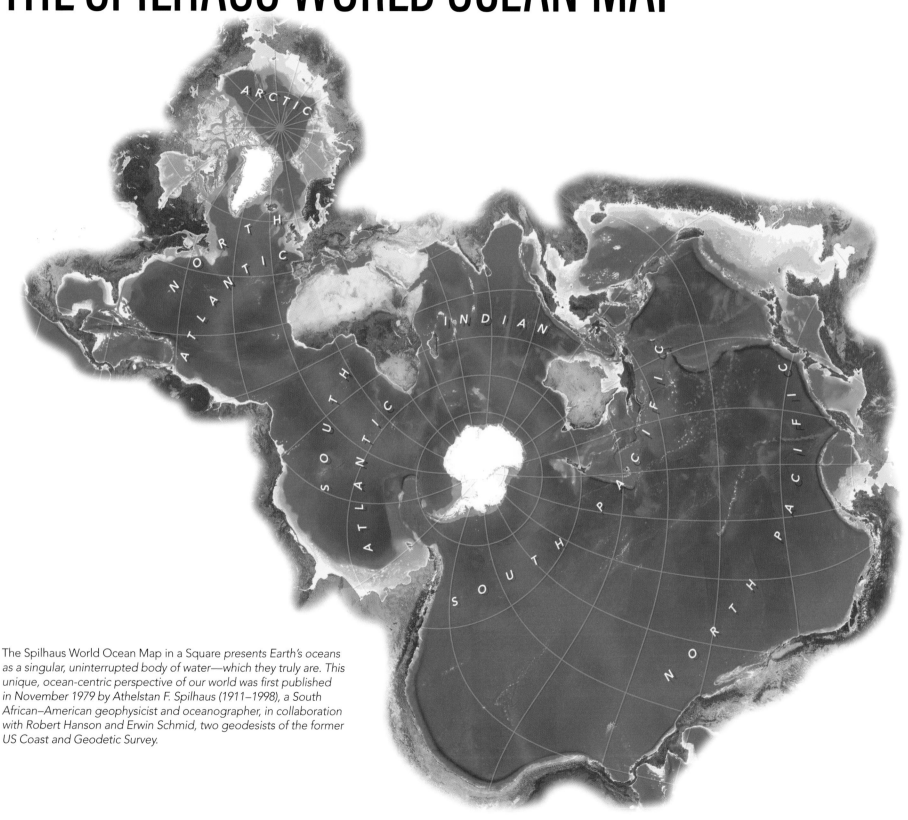

The Spilhaus World Ocean Map in a Square *presents Earth's oceans as a singular, uninterrupted body of water—which they truly are. This unique, ocean-centric perspective of our world was first published in November 1979 by Athelstan F. Spilhaus (1911–1998), a South African–American geophysicist and oceanographer, in collaboration with Robert Hanson and Erwin Schmid, two geodesists of the former US Coast and Geodetic Survey.*

A MAP THAT DEPICTS THE OCEANS AS ONE CONTINUOUS BODY OF WATER

Bojan Šavrič, John Nelson, and David Burrows, Esri

To see the oceans, slice up the land

Most maps portray Earth's surface so that the edge of the map slices some features. Because maps commonly focus on the land and its features, they often portray oceans on the edges and split them. For example, pseudocylindrical world maps centered on the Greenwich Prime Meridian divide the Pacific Ocean into two parts, depicted along the left and right edges. Oceanographer Athelstan F. Spilhaus wanted exactly the opposite, with all the oceans in the middle, sharing a global coastline. To accomplish his goal, projection interruptions must occur over land. For the suitable edge, Spilhaus delineated half of a great circle starting in South China at 115°E and 30°N, ending in Argentina at 65°W and 30°S, and passing near the Bering Strait (which Spilhaus considered to be "so shallow and narrow that it constitutes no real oceanic connection"). These start and end points represent the "poles" projected into the diagonally opposite corners of the projection square and absorbing much of the areal distortion. This edge also interrupts waters in the Sea of Okhotsk, the Bohai Sea, the Gulf of Mexico, near the Gulf of Panama, and along the Peru–Chile Trench. Hence, these areas are repeated at the ocean's perimeter when they are crafted into the Spilhaus World Ocean Map.

An oblique aspect of the Adams Projection of the World in a Square II

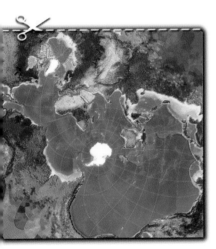

Spilhaus never published exact projection equations for his map, unfortunately. To implement it in GIS software, forward and inverse equations are needed, not only for spherical Earth models but also for ellipsoidal models, such as World Geodetic System (WGS) 1984. Spilhaus created his World Ocean Map in a Square using an oblique aspect of the Adams Projection of the World in a Square II. Oscar S. Adams introduced this projection in 1929, which has remarkable similarities to the Spilhaus map. Both are conformal and portray the world in a square, and both greatly distort areas near the two diagonally opposite corners. Distortions in the other two corners are smaller. Just like the Spilhaus map, the Adams projection can also be mosaicked into an infinitely continuous map of the world.

The Adams projection can be used to reverse-engineer Spilhaus's exact configuration by setting the edge of the map near the Bering Strait. Adams derived the equations for his projection first by conceptually shrinking the world into a hemisphere while maintaining conformality. Then he applied elliptic functions previously used by Charles S. Peirce and Émile Guyou to project the curved surface onto a plane. Adams presented the forward equations for spherical Earth models only (and also only in an equatorial orientation). However, most of today's geospatial data is defined based on ellipsoidal models, such as WGS 1984 or Geodetic Reference System (GRS) 1980. In modern GIS, one also needs the ability to convert projected data back to geographic coordinates, requiring inverse equations. The forward and inverse equations for ellipsoidal Earth models can be achieved by converting geodetic coordinates to a conformal sphere, conformally shrinking the model to a hemisphere, and resolving a complex elliptic integral of the first kind. Esri developed these equations for the Adams Square II projection, available in the latest version of ArcGIS.

The Spilhaus projected coordinate system and the map

The Spilhaus World Ocean Map in a Square or "Spilhaus projection" is also available in the latest ArcGIS software as the WGS 1984 Spilhaus Ocean Map in a Square projected coordinate system. Its well-known ID is 54099. The projection parameters are derived from the edge of the map passing through the same three points used by Spilhaus, starting in South China, passing across the Bering Strait, and ending in Argentina. The only difference is that the edge does not represent a great circle on a sphere but rather a carefully crafted curve on the surface of the WGS 1984 ellipsoid passing exactly through all three points of the edge.

Creating a continuous world ocean map

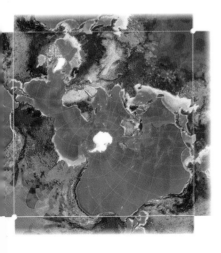

The WGS 1984 Spilhaus Ocean Map in a Square projected coordinate system cannot repeat areas, so a single instance of the map will appear to be clipped along the edge of the Gulf of Mexico and the Bering Strait. A visually pleasing layout without apparent clipped edges can be crafted in ArcGIS in three steps. First, duplicate an already-styled map four times within a layout and position them at each of the four edges of the center map. Second, rotate each of the four perimeter maps such that its coastal edge aligns with its neighbor. The repeated portions of water along the edge provide a sense of continuity. Finally, and optionally, overlay a visual graphic to occlude the overly redundant areas in the layout, or simply position the underlying layout to clip the perimeter maps to your preference. The result is an uninterrupted world ocean map. Understanding its areal distortion, and therefore limitations for thematic mapping, the unique perspective of the Spilhaus World Ocean Map in a Square justifies its use for important, highly visual messages about the largest ecosystem of our planet.

Dashboard and map view of the state of the COVID-19 outbreak as of March 9, 2020.

COVID-19 DASHBOARDS
MONITORING A PANDEMIC

ek Law, Esri

December 2019, an outbreak of a pneumonia-like disease was reported in Wuhan
y, China. The virus was named SARS-CoV-2 (novel coronavirus). By the time the World
alth Organization (WHO) and US Centers for Disease Control (CDC) properly identified
e virus, it had spread to countries beyond China.

early February 2020, Lauren Gardner, director of the Center for Systems Science and
gineering and a civil engineering professor at Johns Hopkins University (JHU), led a team to create
ashboard to help monitor and visualize reported cases on a global scale (see this book's introduction).

e JHU COVID-19 Dashboard reports case locations and key performance indicators, such as confirmed cases,
aths, and recoveries. Confirmed cases can be filtered by country, and, in the United States, further filtered by state
d county. Serial charts show data trends over time.

early March, many other countries began reporting COVID-19 cases. Several hundred different dashboards have been created to
elp monitor the outbreaks. Some are designed for country-specific or, in the United States, at state, county, and local levels. Variations
the dashboards have also been created to support viewing on mobile devices.

learn more about ArcGIS Dashboards and visit the live apps mentioned above visit GISforScience.com.

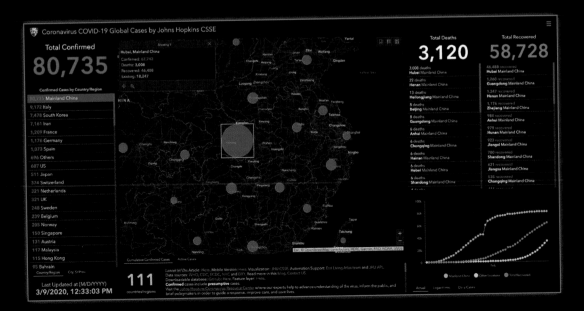

EXPLORING SPATIOTEMPORAL PATTERNS
MAPPING HARMFUL RED TIDE DATA

Ankita Bakshi, Esri

When harmful red tides form off the coast of Florida, the state's Fish and Wildlife Research Institute (FWRI) is there to collect data and document the phenomenon. A red tide, or harmful algal bloom (HAB), occurs when colonies of algae grow in abnormally high concentrations and produce toxins that can harm marine ecosystems and public health. In Florida and the Gulf of Mexico, the species that causes most HAB events is *Karenia brevis*, which in high concentrations discolors water a reddish-brown hue, hence the name *red tide*. By understanding the spatial and temporal dimensions of the red tide observations, the FWRI can implement targeted monitoring research strategies to reduce the cost and improve the efficiency of sampling efforts.

Understanding red tide data spatially

Identifying areas of high values (hot spots) and low values (cold spots) using the ArcGIS Hot Spot Analysis tool is one of the most common ways to start exploring and analyzing data spatially. Using the Getis-Ord Gi* statistic can identify statistically significant clusters of high and low red tide observations. Applied to the FWRI red tide data, this technique found a statistically stronger presence of HAB along the southwestern Florida coast extending all the way to the embayment area in the northwest.

Visualizing the red tide data temporally

The duration of algal blooms varies based on physical and climatic conditions. We can use the data clock temporal chart to understand these patterns and the data in time. A data clock visually summarizes temporal data into 2D and reveals seasonal or cyclical patterns and trends over time. The temporal distribution of red tide in this chart reveals a higher frequency of red tide observations in the months of fall. The blooms mostly occur from September to November. In some years, the bloom's stay was short-lived, but in most years the red tide continued into the winter months.

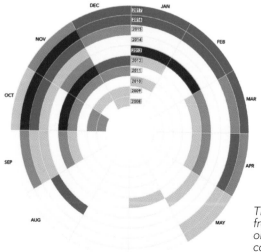

This data clock shows higher frequencies of red tide observations along Florida's coasts from September to November.

Mapping the red tide data in space and time

Powerful 2D and 3D visualization techniques and integrated trend analysis help us visualize and analyze the presence of blooms simultaneously in space and time. The Create Space Time Cube By Aggregating Points tool in ArcGIS Pro summarizes a set of points into a netCDF data structure that can be thought of as a 3D cube made up of space-time bins. A common way to aggregate points spatially is to aggregate them to a regularly shaped grid (either a fishnet or hexagon shape).

3D map (detail) of red tide observations over time along Florida's western coast.

A space-time cube was created by annual aggregation of red tide observations to hexagon bins with a spatial extent of 10 square kilometers. Any missing data, which is common in monitoring and sampling data, were filled with the average value of space-time neighbors. The map shows a 3D visualization of the space-time cube in which each bin represents the number of red tide observations in a year for each location. ArcGIS Pro includes many display options, all with preset symbology and range and time sliders that make the exploration of the space-time cube and analysis results intuitive.

To quantify and understand the patterns in these thousands of stacked bins, we can visualize the cube in 2D and calculate trend analysis. Trends analysis in the Visualize Space Time Cube in 2D tool shows where the red tide observations have increased or decreased over time using the Mann-Kendall statistic. The dark-green locations show a downward trend in the observations. The locations in dark purple have an upward trend of red tide observations with 99 percent statistical confidence. The results are consistent with the spatial analysis using the Hot Spot Analysis tool. The areas in southwestern Florida near Tampa Bay mostly show an upward trend. The embayment areas behind the barrier islands such as the Apalachicola Bay and St. George Sound, where the tributaries drain nutrient-rich water, also show an upward trend.

Using visualizations and analysis in space, time, and space-time helps us better understand the patterns of red tide in Florida's coastal waters. These visualizations are just a subset of many methods you can use to explore and analyze space-time data in ArcGIS Pro.

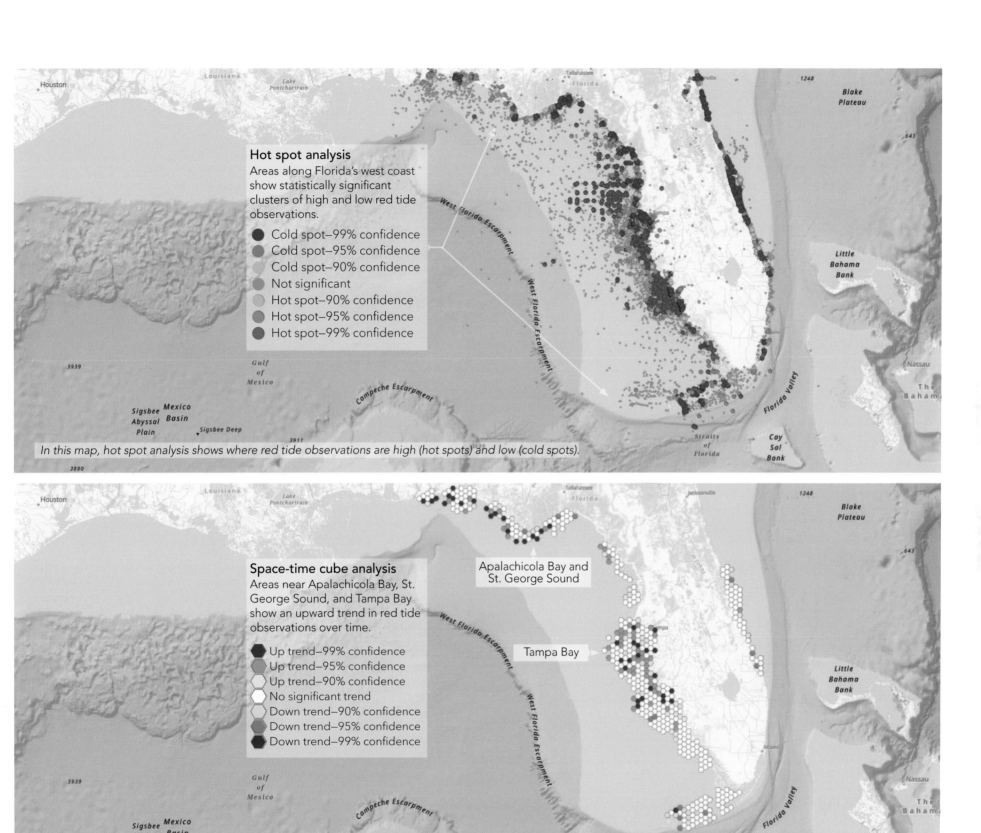

Hot spot analysis

Areas along Florida's west coast show statistically significant clusters of high and low red tide observations.

- ● Cold spot–99% confidence
- ● Cold spot–95% confidence
- ● Cold spot–90% confidence
- ● Not significant
- ● Hot spot–90% confidence
- ● Hot spot–95% confidence
- ● Hot spot–99% confidence

In this map, hot spot analysis shows where red tide observations are high (hot spots) and low (cold spots).

Space-time cube analysis

Areas near Apalachicola Bay, St. George Sound, and Tampa Bay show an upward trend in red tide observations over time.

- ⬡ Up trend–99% confidence
- ⬡ Up trend–95% confidence
- ⬡ Up trend–90% confidence
- ⬡ No significant trend
- ⬡ Down trend–90% confidence
- ⬡ Down trend–95% confidence
- ⬡ Down trend–99% confidence

Apalachicola Bay and St. George Sound

Tampa Bay

In this map, trend analysis shows where red tide observations have increased or decreased over time.

MODELING RELATIONSHIPS USING LINK ANALYSIS

EBOLA OUTBREAK, SIERRA LEONE 2014 - 2015

Linda Beale, Esri

Link analysis played a strong role in tracking the person-to-person transmission of the Ebola virus in Sierra Leone in 2014–2015. Link analysis uses graph theory for evaluating connections or relationships between nodes, where nodes can represent people, places, objects, and events. You can visualize the results of link analysis using an association matrix, or more typically, a link chart to evaluate the patterns of interest. Geographically, flow maps are used to show the movement of objects from one location to another.

Link analysis

Several different measures of topological centrality are possible with link analysis, each of which seeks to answer a slightly different question. The degree of nodes shows the measure of centrality, and normalized centrality measures adjust for network size.

Degree centrality allows you to see what is flowing through the network and identify the most influential nodes. The important nodes are identified as those having the most connections. Degree centrality can have directionality so that nodes with higher out-degree values are more central, or nodes with higher in-degree are more important. Degree centrality is a local measure that considers a node's importance within its locality, but not any indirect relationships.

Betweenness centrality measures the extent to which a node lies on paths between other nodes. Nodes with high betweenness are likely to have an important influence within a network by virtue of their control over information passing between other nodes. Removal of nodes with high betweenness from the network will have the greatest disruption on communications or flow across that network as they lie on the largest number of paths.

Closeness centrality is based on the average of the shortest network path distance between nodes and identifies nodes as being more central if they are closer to most of the nodes in the network. Closeness centrality is used to determine which nodes are most closely associated to the other nodes in the network.

Eigenvector centrality depends on the number of neighbors and the quality of its connections, with the most central nodes being important nodes that are connected to other important nodes. Eigenvector centrality is of value to determine the nodes that are part of a cluster of influence.

Link analysis together with spatial data analysis offer enormous value for epidemiological analysis of distributions, patterns, and determinants of health and disease conditions within populations. Understanding the development of epidemics

caused by infectious diseases and the impact of interventions, together with an understanding of the geography of at-risk populations and potential transmission pathways, can help ensure effective responses in the future.

Pathways of transmission

Infectious disease epidemiology can use link analysis to show connectivity of individuals or places. The measure of centrality allows the isolation or accessibility to be measured. If a link directly connects two nodes, these nodes can be evaluated as transmission events from individual to individual or place to place. These relationships indicate potential transmission pathways for infections between individuals or through populations.

Interactions between microorganisms such as bacteria and viruses cause infectious diseases. Zoonotic diseases are infectious diseases of animals that can cause disease when transmitted to humans.

A transmission network can be created using individual data of infected people linked to those from whom they caught the infection and to any others they infected. This network will show all the links through which infection spread in the outbreak; however, it will not show interactions that led to infection transmission. Because nodes represent places at a population level, the nodes represent locations of high connectivity of infected cases, which together with population data can help define those areas where population interactions were highest.

West Africa, Ebola outbreak 2014–2015

Data from the World Health Organization shows Ebola cases in Sierra Leone from the 2014–2015 Ebola outbreak in West Africa. Ebola, a zoonotic disease, spreads in the human population through human-to-human transmission. Home, infection, and death locations for known cases show the geographic spread, with the node sizes showing the degree centrality value of those locations.

Research shows that during the 2014–2015 West Africa outbreak, the majority of transmission events occurred between family members. The link chart shows the relationships to known contacts with people diagnosed with Ebola. Understanding traditional practices and Ebola transmission pathways ultimately led to changes in behaviors related to mourning and the adoption of safe burial practices.

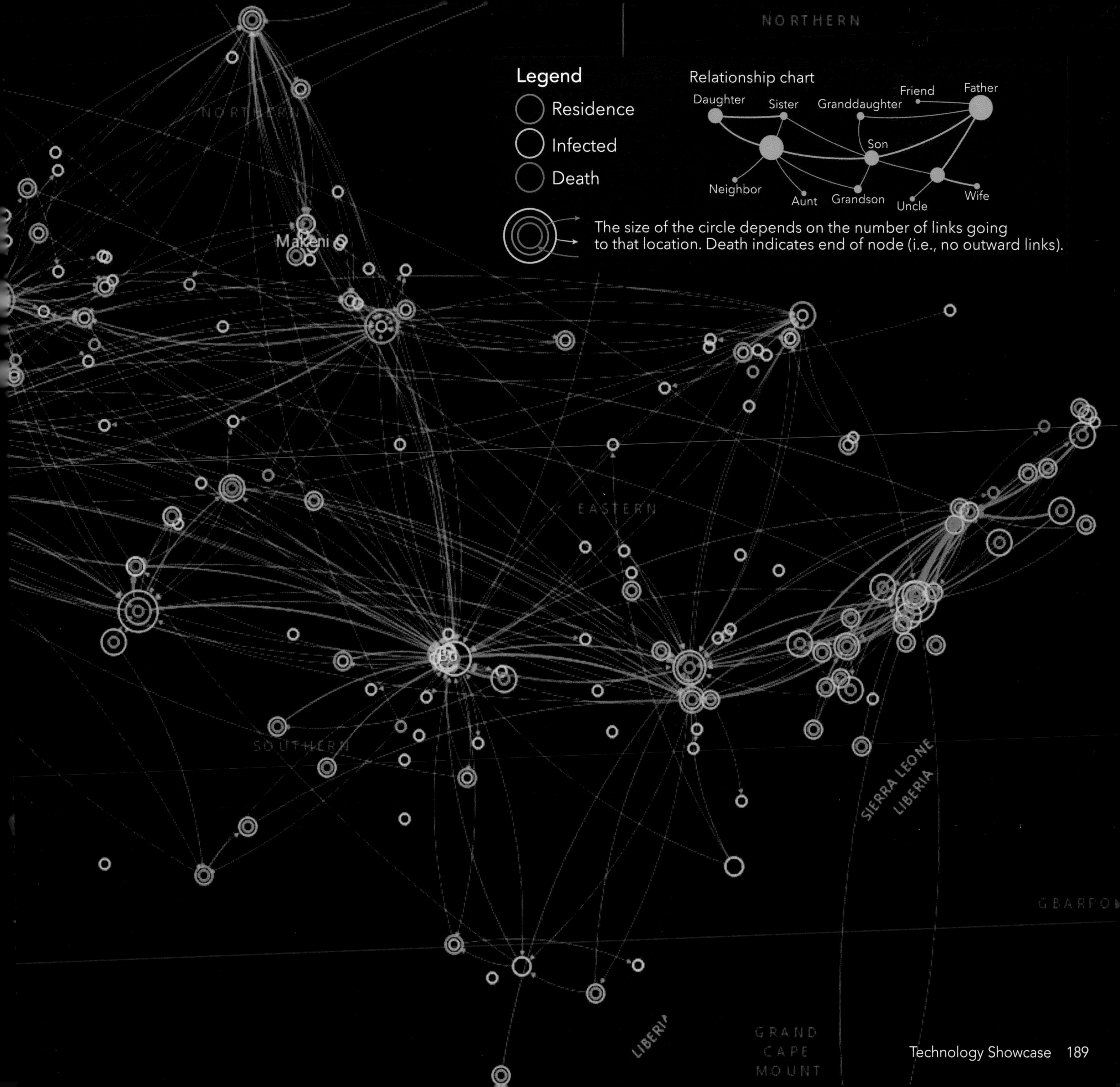

NORTHERN
NORTHERN
Makeni
EASTERN
SOUTHERN
Bo
SIERRA LEONE
LIBERIA
LIBERIA
GRAND CAPE MOUNT
GBARPO
Legend
Residence
Infected
Death
Relationship chart
Daughter
Sister
Granddaughter
Friend
Father
Son
Neighbor
Aunt
Grandson
Uncle
Wife
The size of the circle depends on the number of links going to that location. Death indicates end of node (i.e., no outward links).

SMART MAPPING AND ARCADE
TRANSFORM YOUR THEMATIC THINKING

Mark Harrower and Jim Herries, Esri

ArcGIS® Arcade is a simple scripting language for data in your maps. Arcade frees you to explore anyone's published data, no download required, saving valuable hours of time. The expressions created in Arcade run on the fly, meaning you do not have to own the layer to calculate what you need for your map style, pop-up, or label.

Smart mapping is built into ArcGIS Online as a simple workflow for exploring your data, trying various styles of maps that might suit your data, and polishing the map's final appearance. Smart mapping takes the guesswork out of making great thematic maps by using data-driven styling and intelligent defaults. But how do these two capabilities in ArcGIS change how you understand and map your data?

Arcade lets you explore and even extend the data without having to own it on a hard drive. Working with the data, you begin to see what you need for your map's symbols, pop-ups, and labels.

Arcade expressions can be as simple as calculating a percentage, or converting a year from a string to a number. Other expressions use if-then logic to look at several attribute fields in each record and return an evaluation of that record's data. Many of the map styles available in smart mapping are rooted in Arcade expressions that find relationships and patterns hidden among the columns of attribute data.

Crafting good maps traditionally requires authors to make dozens of interrelated decisions: what renderers, scales, basemaps, and colors to use; how many labels to use; how to classify the data; and whether the map needs boundaries, and if so, how to draw them.

Smart mapping taps into that desire for an easier but also useful pathway. Once you choose one or more attributes or Arcade expressions to map, smart mapping examines the types of attributes you chose (text, numbers, date) and suggests map styles for you to use. It doesn't force you to use these settings. You can override the settings as needed.

Iteration and exploration are founding principles of geographic visualization. Why? Because data are complex and no single depiction of the data can answer every question. Sometimes we start a project thinking we know what kind of map or treatment we need, only to find, after some exploration, that there are more productive avenues to explore. Using Arcade makes it easier to uncover new patterns and share hidden insights within your data.

After settling on a map style, you then bring your expertise or research into play. The best maps use some kind of standard of comparison to communicate what's "normal" and what's not. You change the default colors and sizes to emphasize what's important and de-emphasize what is less important—just as a road map clearly distinguishes major highways from residential streets and filters out dirt roads altogether.

Let's consider some real examples. Say you are a climate scientist and want to compare average air temperatures for two different years for a bunch of cities. You could make one map for each year and scan back and forth to try and spot differences. But Arcade allows you to calculate the difference in temperature between two years—on the fly—and makes it easy to see which cities are getting warmer or cooler. The ability to derive new data from the raw data using Arcade expressions is the real magic in cartographic analysis.

Flexible data manipulation tools joined with flexible map visualization tools allow us to work at a pace unheard of in the past. You can say "show me only events that happened on Mondays" or "remove areas larger than 100 sq km and rerun" or "proportionally increase the symbol size of stores based on % year over year sales growth, because we need to see regional sales growth trends."

Arcade can also help connect people to your maps by changing data into information and insight. You can easily format facts in map pop-ups so they are easier to digest. Earlier mapmaking technology often required users to read pop-ups with capitalized field names that seemed to shout at the reader ("AVGTMP02: 65.66"). But testing shows people learn and recall better if that information is presented in the same way that we speak ("In July 2002, the average temperature in London was 65.7°F"). But why stop there? Facts are far more meaningful when they're contextualized. Arcade allows us to automatically derive context and significance for our information: "In July 2002, the average temperature in London was 65.7°F, while in 2019, it was 68.9°F (3.1 degrees higher)." Explaining information this way allows readers to uncover insights without having to do all the busywork.

The old pattern was to dump the data onscreen and let readers sort it out, if possible. Useful pop-ups verbally reinforce the map patterns that smart mapping reveals. The goal of all useful and engaging maps—whether they're made for an audience of one or one million—is to turn data into insights and understanding.

In Eagle River Union Airport, a decrease of -6 degrees (F) is expected in the next 24 hours at station KEGV.

Accuweather information as of 3/2/2020, 8:57 AM

Zoom to   Get Directions

Temperature departure in 24 hours

Red crosses are increases in temperature. The higher the increase, the more they lean to the right. Blue crosses indicate cooling temperatures; the more they lean to the left, the bigger the cooling trend.

< -20   0   > 20

Data provided by Accuweather.

DEEP LEARNING IN DISASTER MANAGEMENT

The 2018 Woolsey Fire burned in Los Angeles and Ventura Counties, California. Responders used deep learning with GIS to quickly perform feature extraction from imagery to identify damaged and undamaged buildings. This analysis enabled first responders to find damaged and destroyed structures and deploy the appropriate help and resources.

2018 Woolsey Fire

Fire perimeter
Damaged building
Undamaged building
Imagery by DataWing Global.

IMAGERY FEATURE EXTRACTION TO IDENTIFY DAMAGED STRUCTURES

Ling Tang and Simon Woo, Esri

Wildfires can spread quickly and destroy thousands of acres of land. In California, where many urban areas infringe on shrublands, these fires often threaten homes and lives. The 2018 Woolsey Fire—a notably fierce and fast-moving blaze—killed three people, destroyed more than 1,600 structures, forced nearly 300,000 residents from their homes, and burned in total nearly 100,000 acres.

During a disaster, time is essential for response and recovery efforts. Slow response times can result from the lack of timely data and manual interpretation of this data. In these situations, first responders must quickly and accurately identify urban structures that fires have destroyed or otherwise spared. This information is vital to first responders, government agencies, and insurance adjusters. The use of drone imagery and Esri's deep learning tools efficiently processes and analyzes imagery for timely decision support.

As firefighters extinguished flames, drones flew over the burned areas to assess damage. DataWing Global, an aerial data services company and Esri partner, captured 40 gigabytes of on-demand, high-resolution imagery over the Woolsey area. To manage this vast amount of data, a mosaic dataset was created, which allowed responders to display, analyze, and share a collection of images.

Once the data is managed within a mosaic dataset, users can perform deep learning analysis. Deep learning is a subset of machine learning, in which learning is based on an algorithm known as an artificial neural network. Artificial neural networks are computing systems that recognize and learn patterns. Training and deploying a deep learning model involves three steps: creating training samples, training the deep learning model, and running the model inference, which yields a classified map of features.

First, training samples must be created to categorize the damaged and undamaged structures. Building footprints from the Los Angeles County GIS Data Portal were draped over the orthorectified, high-resolution drone imagery, and a new "ClassValue" field was added to the building footprint feature class. Firefighters used this field to identify and label buildings as "damaged" or "undamaged." These categorized features were exported using the Export Training Data for Deep Learning tool in ArcGIS® Pro. Training a deep learning model in ArcGIS Pro requires users to set up the Python environment with the necessary deep learning libraries, including PyTorch, Fastai, and library dependencies.

The Train Deep Learning Model tool in ArcGIS Pro used the labeled training samples to train a building damage classification model. The model type is preconfigured as "Feature Classifier" based on the metadata format of the training samples. This geoprocessing tool calls the third-party deep learning application programming interfaces (APIs)—like PyTorch or Fastai—to perform the model training tasks. This tool provides optimal model training parameters for training the damage classification model. The model was trained using a ResNet architecture to classify all buildings in the imagery as either damaged or undamaged. During the model training process, messages regarding training loss, validation loss, and accuracy are generated after each training step. This process allows users to monitor the training progress.

Once the training was complete, the manually assigned ground-truth labels were compared to the model classification results to assess model performance. The results show that the accuracy rate of identifying damaged and undamaged structures was more than 99 percent. The saved model includes the model binary file and the Esri model definition (.emd) file, which can be used to perform model inference in ArcGIS Pro. A zipped deep learning model package (DLPK) file can be shared on the ArcGIS Portal and deployed in ArcGIS Enterprise. Raster analytics (RA) tools can use a DLPK, an item type in Portal for ArcGIS. A DLPK is a compressed file, portable, and easy to use and share.

The Classify Objects Using Deep Learning tool in ArcGIS Pro was used to perform model inference and classify the buildings. Both postdisaster imagery and the building footprint feature class were used as inputs. The result is an updated feature class of the building footprints, with a new ClassLabel field to assign each building as either damaged or undamaged. By running inferencing inside ArcGIS Enterprise using the model and classify objects function in ArcGIS API for Python arcgis.learn module, inferencing can be scaled for large projects by leveraging the RA capability on ArcGIS Image Server.

More than 9,000 buildings were automatically classified. Of those, more than 1,300 buildings were deemed as damaged or destroyed by the fire. The resulting map shows the damaged buildings as red and the undamaged buildings as green. With a 99 percent accuracy rate, the deep learning model is as accurate as a trained adjuster and much faster. What usually takes a week was performed in a few hours.

Using the ArcGIS Infographics Add-In, first responders combined analysis with demographic data to further identify at-risk populations, such as children and the elderly. The Infographics report can be generated to quickly provide statistics and other information to assess the magnitude of the situation and help deploy the proper help.

This report was created with the ArcGIS Infographics Add-In.

SEA-SURFACE TEMPERATURE TREND MAPPING

Annual SST change over time
(1981–2019)

Increase   Decrease

Sea-surface temperature trend raster map (1981–2019).

ANALYZING TIME-SERIES DATA

Hong Xu, Esri

More than 70 percent of Earth's surface is ocean, so sea-surface temperature plays a major role in regulating Earth's climate system and serves as an important indicator of climate change. The ocean absorbs vast quantities of heat from greenhouse gas emissions, leading to rising ocean temperatures and changing ocean circulation patterns that transport warm and cold water around the globe. The National Oceanic and Atmospheric Administration (NOAA) reports that the average global sea-surface temperature has increased by approximately 0.13 degrees Celsius (32.23 degrees Fahrenheit) per decade during the past 100 years. Increasing sea-surface temperatures will substantially affect climate, marine species, and ecosystems. Researchers predict that rising temperatures are already contributing to species extinctions and extirpations, rising seas, and flooded ecosystems, for example.

Modern remote sensing technology and high-resolution time-series data help scientists study changing sea-surface temperatures. They use geographic information systems (GIS) and statistical regression methods to better understand spatial variations in sea temperature change over time.

The method of modeling annual and seasonal trends

Many variables in Earth science exhibit periodicity. For example, temperatures are high in summer and low in winter each year. You can use a linear regression to model the general trends of recurring phenomena, but linear regression is not suitable to describe the seasonality of the variables. Scientists use a method called *harmonic regression*, which tends to exhibit periodic rhythms, to model annual long-term trends and seasonal changes over time.

The Generate Trend Raster tool in ArcGIS integrates statistical regression methods into the multidimensional raster data model. The multidimensional raster contains multiple rasters representing data at different times (or other dimensions). The tool computes a regression model using the harmonic algorithm for each pixel array along time and outputs a trend raster that contains the regression models of each time series. The regression model coefficients and the statistical terms such as root mean square error (RMSE), p-value, and R-squared are stored as bands in the output trend raster,

Predicted monthly SST, 2050

Highest — Lowest

Masked pixels

Map of predicted sea-surface temperature (SST) by 2050.

which can be used to visualize seasonal trend, map the annual trend, evaluate model performance, and predict future data. The Generate Trend Raster tool and Predict Using Trend Raster tool built based on harmonic regression will be used in analyzing sea-surface temperature.

Data preparation

The data used in this analysis are from NOAA's daily sea-surface temperature (SST) data (1981–2019) with a spatial resolution of 0.25 × 0.25 degrees. The data consists of 39 Network Common Data Form (NetCDF) files (one file per year) and represent 13,931 images. First, a mosaic dataset is created from the NetCDF files. Next, the mosaic dataset is converted to a multidimensional cloud raster format (CRF), which stores multidimensional rasters for optimal time-series image analysis and multidimensional raster computing. Next, daily SST data was aggregated into monthly SST data by averaging the pixel values of each month. Finally, analysts built a transpose for the multidimensional CRF to speed up the across-time dimension computing.

The sea-surface temperature trend map

The team created a trend raster using the Generate Trend Raster tool. In the output trend raster, the band named Slope is used to map the long-term trend as shown, where positive (purple) indicates sea-surface temperature increases annually with time while a negative value indicates a decrease (green).

The trend map shows that the sea-surface temperature changes dynamically in different parts of the ocean. While most of the domain exhibits an increase trend, the increase varies spatially. The darker purple indicates a greater increase rate. For example, some regions of the North Atlantic show greater increases than in the North Pacific. The long-term annual trend at a location of the North Atlantic Ocean has a slope of 0.00008, and the seasonal trend at that location is also clearly modeled by the harmonic regression. The ocean in the Southern Hemisphere also shows temperature increase in general except the decrease trend in the East Pacific Ocean off the coast of South America, where the occurrences of El Niño and La Niña cause the temperature to change dramatically and cause a negative slope in the harmonic model. Analysis of the model's accuracy using the R-squared band of the trend raster shows that this regression model fits well overall except in some areas close to the equator or polar regions, where sea-surface temperature does not have obvious seasonal effect.

Finally, a layer was generated from the R-squared band using raster function to select and mask out pixels with less model accuracy (R-squared < 0.6). The Predict tool was used to generate a predicted SST of the next 30 years from the trend raster. Calculated from the predicted SST, the average SST will increase from 31.57 in 2019 to 32.01 by 2050.

Data source: NOAA/OAR/ESRL PSD, www.esrl.noaa.gov/psd.

SPACE-TIME PATTERN MINING
EMERGING HOT SPOT ANALYSIS OF POLLUTION DATA

Lynne Buie, Esri

From massive wildfires that darken the skies over a continent to volcanic eruptions that leave big cities coated in ash, we regularly hear about—and sometimes experience—major pollution events. However, it's not always easy to evaluate whether the pollution from these events is actually worse than anywhere else. The news media may over-report the news in a particular region and largely ignore what happens in another. Or a singular event may leave the impression that an area is heavily polluted when in reality, it may have clean air most of the time. What we do know is that pollution is harmful to human health. Pollution often contains microscopic particulate matter PM2.5, inhalable particles 2.5 micrometers or smaller in size, or about 30 times smaller than the width of a strand of human hair. Elevated levels of PM2.5 have been linked with increased infant mortality, and cardiovascular and pulmonary diseases such as asthma, lung fibrosis, and hardening of the arteries. Aside from the human costs, these and other pollution impacts from PM2.5 have an estimated financial impact of $225 billion per year globally in lost labor.

To get a clearer picture of pollution events, we can use data from Earth observation satellites to measure atmospheric and surface phenomena. One of those measurements is called *aerosol optical depth* (AOD), where the absorption or scattering of light in the atmosphere serves as a proxy for the presence and quantity of PM2.5. Van Donnkelaar et al. (2018) collected the annual AOD from three Earth observation satellites from 1998 to 2016. The researchers combined and refined the data using a chemical transport model and geographically weighted regression (GWR) with ground-based PM2.5 observations.

The global time series of annual PM2.5 pollution data across 19 years is a perfect candidate for space-time pattern mining using ArcGIS Pro, which applies rigorous statistical tests to space and time data to find statistically significant patterns. The basis of the space-time pattern mining analysis is the space-time cube data structure—a method of representing temporal data as a multidimensional array appropriate for analysis.

With the PM2.5 pollution data in a space-time cube, researchers can apply the space-time pattern mining

technique called *emerging hot spot analysis*. This analysis allows researchers to objectively assess areas of high and low pollution by finding hot and cold spots in the global pollution data. A hot (or cold) spot is an area with high (or low) values of pollution and surrounded by other areas of high (or low) values. Researchers assess each location in each time slice independently using the Getis-Ord Gi* statistic to determine whether the time slices are statistically significant hot or cold spots. To incorporate the temporal component, researchers then apply the Mann-Kendall trend test to each location to assess the trend of hot or cold spots. This test results in nine different types of hot and cold spots, depending on the pattern of each location through time.

The results of applying emerging hot spot analysis show that when compared to the rest of the world, much of Southeast Asia and Sub-saharan Africa are hot spots of pollution (shown in shades of red). Cold spots (shown in blue) are found in North Africa, Australia, some coastal parts of South America, and many of the most northern latitudes. The map legend shows the nine different types of hot and cold spots. Persistent hot spots, seen for much of Asia and Africa, have been a statistically significant hot spot for 90 percent of the time-step intervals, with no discernible trend indicating an increase or decrease in the intensity of clustering over time. New hot spots, seen in limited areas of Asia, Africa, and South America, are locations that have statistically significant hot spots for the final year in 2016 and have never been a statistically significant hot spot before.

Sporadic cold spots, seen in large areas of Alaska and northeastern Russia but also in small quantities worldwide, are locations that are on-again, off-again cold spots. Diminishing cold spots, again seen worldwide but particularly in South America and Australia, are locations that have been a statistically significant cold spot for 90 percent of the time-step intervals, including the final year. In addition, the intensity of clustering of low counts in each time step

is decreasing overall, and that decrease is statistically significant for diminishing cold spots. These four types of hot and cold spots, and all the others seen in the legend, help us understand differences in the global pollution patterns through time. Applying emerging hot spot analysis is an effective way to objectively understand space-time patterns in scientific data.

Pacific Ocean

New hot spot
Consecutive hot spot
Intensifying hot spot
Persistent hot spot
Diminishing hot spot
Sporadic hot spot
Oscillating hot spot
Historical hot spot
New cold spot
Consecutive cold spot
Intensifying cold spot
Persistent cold spot
Diminishing cold spot
Sporadic cold spot
Oscillating cold spot
Historical cold spot
No pattern detected

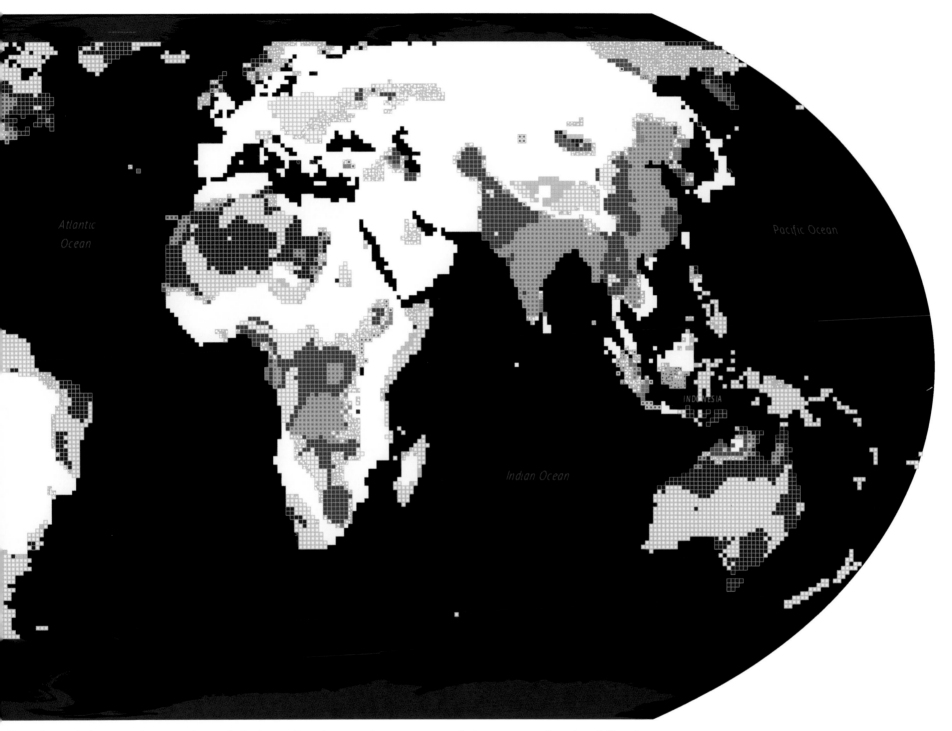

Using this analysis, we can better understand whether regions that experience extreme pollution events, such as the wildfires that raged across Australia starting in late 2019, have worse pollution than regions that may receive less attention and analysis for one reason or another.

Reference: van Donkelaar, A., R. V. Martin, M. Brauer, N. C. Hsu, R. A. Kahn, R. C. Levy, A. Lyapustin, A. M. Sayer, and D. M. Winker. 2018. *Global Annual PM2.5 Grids from MODIS, MISR and SeaWiFS Aerosol Optical Depth (AOD) with GWR, 1998–2016.* Palisades, New York: NASA. Data sources: NASA Socioeconomic Data and Applications, Center Global Annual PM2.5 Grids from MODIS, MISR, and SeaWiFS Aerosol Optical Depth (AOD) with GWR, v1 (1998–2016).

TIME IN SPATIAL SCIENTIFIC WORKFLOWS

Time-series clusters show counties with similar patterns of population growth or decline from 1969 to 2018. Three distinct patterns emerge. The areas shown in red include the Great Plains and Mississippi Delta, where declining populations result partly from the transition from small, family-based farms to large, mechanized agrobusiness. These areas also include parts of Appalachia, where the decline of the coal industry and low birth rates contribute to net migration loss. The areas shown in green include a majority of counties where population has steadily increased. In the areas shown in beige, the population has remained stable throughout the period. A small number of counties shown in gray either had significant changes in their boundaries or were missing data and thus were excluded from the analysis.

TIME-SERIES CLUSTERING OF POPULATION GROWTH AND DECLINE

Kevin Butler, Esri

Highly dynamic Earth changes on temporal scales range from minutes in the case of earthquakes to decades in the case of deforestation. To fully understand our planet, scientists must discern patterns across space—a powerful capability of GIS—and across time. Modern geographical information systems increasingly integrate new methods and techniques for analyzing temporal data. One important aspect of analyzing temporal data is detecting and quantifying patterns. Do different locations have similar patterns of an observed or modeled variable across time? One method for exploring patterns across time is time-series clustering.

Time-series clustering

Time-series clustering partitions a collection of time series based on the similarity of time-series characteristics. In the context of GIS, the collection of time series comes from individual time series at different locations in space. Time series can be clustered so they have similar values in time or similar behaviors or profiles across time (increase or decrease at the same points in time). The ArcGIS Time Series Clustering tool identifies locations that are most similar and partitions them into distinct clusters, where members of each cluster have similar time-series characteristics.

The goal of clustering is to partition the locations into groups where the time series within each group are more similar to each other than they are to the time series outside the group. However, time series are composed of many numbers or values across time, so it is not completely clear what it means for two time series to be similar. For individual numbers, a useful measure of similarity is the absolute difference in their value. For example, the difference between 10 and 13 is 3. You can say that 10 is more similar to 13 than it is to 17 because the absolute difference in their values is smaller. For time series, however, the similarity is less obvious. For example, is the time series $(5, 8, 11, 7, 6)$ more similar to $(4, 9, 13, 4, 9)$ than it is to $(5, 11, 6, 7, 6)$? To answer this question, you must measure how similar or different two time series are. Each of the several ways to measure similarity depends on which characteristics of the time series you consider important. You can cluster time series based on the raw values of the time series, the correlation between time series, or the shapes of cyclical patterns in the time series.

Increasing similarity

When you cluster based on raw values, the similarity between time series is quantified by the sum of the squared differences in value across time (Euclidean distance in data-space). When you cluster based on correlation, time series are considered similar if they tend to stay in consistent proportion with each other and increase and decrease in value at the same time. To cluster time series that have similar smooth, periodic patterns in their values across time, the time series are decomposed into basis functions from the Fourier family and are represented by oscillating sine and cosine functions with varying periods; these periods are sometimes called *cycles* or *seasons*. Time series are considered similar if the periods of their dominant basis functions are similar. All three methods return a single number that measures the difference between two time series. This difference is calculated for every pair of locations in the study area and is summarized as a dissimilarity matrix. This matrix is then clustered using the k-medoids algorithm. This algorithm finds clusters within the matrix in which members of the clusters are more similar than members of other clusters. This algorithm is random in nature, and it works by choosing random locations to serve as representatives of each cluster. These representatives are called *medoids*, which are analogous to the median of a univariate dataset. Initial clusters are created by assigning every other location to the cluster whose medoid is most similar. The algorithm then swaps medoids within each cluster and reevaluates the similarity within the new clusters. If the new clusters are more similar than the initial clusters, the medoids are swapped, and the process repeats until there are no swaps that will increase the similarity of the clusters.

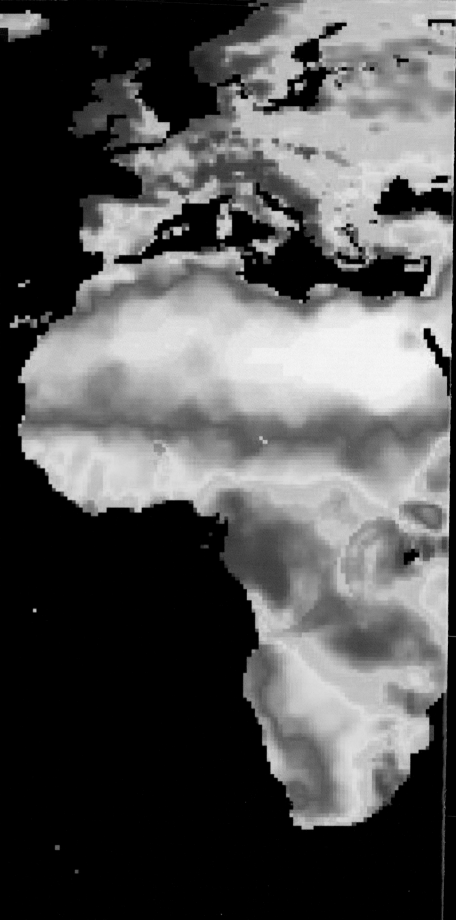

ACCESSING SCIENTIFIC DATA IN THE CLOUD
MODELING GLOBAL SOIL MOISTURE

Kevin Butler, Esri

The practice of science has changed. Scientists increasingly acquire data instead of just directly measuring or observing the data. While independent field observations continue to be important, the familiar leather-bound field notebook has given way to massive central repositories of scientific data often remotely sensed by satellites, automated cameras, autonomous buoys, and drones. The availability of these massive repositories has impacted the spatial scale at which scientists work. These changes in the practice of science are particularly noticeable in the domains of the atmospheric, oceanic, and solid earth sciences.

OPeNDAP and THREDDS

Open-source Project for a Network Data Access Protocol (OPeNDAP) makes data stored on a remote server accessible to you locally, in the format you need, regardless of its format on the remote server. Many authoritative data providers, such as the National Oceanic and Atmospheric Administration (NOAA) and NASA, provide their data product through OPeNDAP data servers. A key value of the OPeNDAP approach is its ability to pull data subsets from the server to get only the data that is relevant to you. ArcGIS provides support for OPeNDAP through the Make OPeNDAP Raster Layer tool.

The University Corporation for Atmospheric Research (UCAR) has created and freely distributes a web server specifically designed for the dissemination of scientific data. The Thematic Realtime Environmental Distributed Data Services (THREDDS) Data Server (TDS) is a web server that provides metadata, web-based catalogs of data and data access protocols for scientific datasets. In addition to OPeNDAP, THREDDS can deliver data as OGC WMS and WCS services, HTTP, and other remote data access protocols.

Storing data in NetCDF

Conceptually, Network Common Data Form (NetCDF) stores the data as multidimensional arrays. Intuitive arrays of data enable efficient access to data along different dimensions. For example, using the same dataset, you may want to draw a 2D map of temperature at a particular altitude and time or create a line graph of temperature values through time at a single location for a specific altitude. In the netCDF file, the data would be represented as a 4D array: temperature (x, y, altitude, time).

The Open Geospatial Consortium (OGC) has adopted NetCDF as a core encoding standard. You can store any type of spatial data in a netCDF file, including atmospheric and oceanic sciences data. The netCDF has the major benefit of containing metadata information and a standard way to describe what each variable represents, its measurement units, and the spatial and temporal properties of the data.

ArcGIS Pro provides a set of tools to represent observations and models as data tables, points, raster fields, or multidimensional raster layers. These tools read netCDF files and format their contents into the corresponding GIS structures, including animations and time-series analysis.

This map shows the long-term monthly mean soil moisture for January. The data used to model this phenomenon is from NOAA's OPeNDAP server. Soil moisture estimates play an important role in long-range temperature forecasts and hydrological studies.

Mean soil moisture for January

Low   High

DRONES AND CITIZEN SCIENCE
THE ADVANTAGE OF COMMUNITY-BASED RESEARCH

Charmel Menzel and Lain Graham, Esri; and Timothy L. Hawthorne, PhD, University of Central Florida

Do you need to capture high-quality, time-sensitive data? Every researcher we know would say, "Of course!" To meet the data and analysis needs required for research, scientists are harnessing the power of drone technology and everyday citizen scientists. In this case, researchers and citizen scientists collected drone imagery using DJI apps for route planning and ArcGIS Drone2Map to convert drone data into high-resolution imagery products while in the field. Collaborative geospatial technologies allow people who don't own drones to participate in the project, share their time and local knowledge, and support the research.

A team from Citizen Science GIS, an international research organization at the University of Central Florida, has worked with citizens in Belize to support data collection and use in various communities. The team uses spatial thinking, interdisciplinary and community-based approaches, Esri products, and drone technologies[2] to (1) make science more accessible and (2) ensure that society informs science and also benefits from scientific discoveries.

Drone imagery collection

The use of drones has made remote sensing available as a personal technology. Esri technology allows users to collect, process, analyze, and share drone imagery. Drone imagery provides current, high-resolution basemaps and can support change detection, feature identification, classification, and analysis. Since 2016, the Citizen Science GIS team and community partners have captured drone imagery of Hopkins Village in Belize annually, extending the flight plans each year to collect data in areas of local interest.

As more people undertake training and get certified as drone pilots, the quality of the data is increasingly reliable. In-field processing tools within ArcGIS Drone2Map ensure quality data capture, including verification that the area coverage and desired accuracy have been achieved. The last thing a project leader wants is to return to the office from the field hundreds or thousands of miles away and learn that the data is not adequate. In this case, while in Belize, a 2D orthomosaic was created from the drone images and shared with citizens. Community members provided valuable feedback during the review process, suggesting that the drone imagery collection add additional areas outside the original flight plan, which are important for context and the village's ecosystem. After reviewing the high-resolution drone imagery, participants recommended additional analysis that previously had not been considered.

Citizen scientist participation

Now that drone imagery has been captured, how does the community use the data? Storm surge, sea-level rise, and development contribute to the flood risk that threatens Hopkins Village, which is located between the Caribbean Sea on the east and a lagoon on the west. So the community needed to better understand its flooding risk and vulnerability with additional data based on the drone-imagery. Prior to drone-imagery data collection, the village did not have a current and reliable GIS dataset of the coastal area, community structures, and road networks. The drone imagery provided a base to create feature datasets. Citizen scientists and students received training to capture data about each structure digitized from the accurate drone imagery, including building material, roof type, use of structure, number of floors, and elevation. They used structure data to calculate vulnerability for each building as described in the journal article, integrating sketch mapping and hot spots analysis to enhance capacity for community-level flood and disaster risk management.

The involvement of community members helped them better understand the data and use the information to support decision-making. In the spirit of open science, the public, community leaders, and researchers can access the open data portal, which includes drone imagery and basic data about culverts and drainage, flooding, and street networks hosted on ArcGIS Online. These datasets differ from most of the larger proprietary or government-controlled datasets in that the local community helped create them. The annual collection of drone imagery resulted in more efficient citizen-led projects such as coastal debris cleanup. Additional research projects include analyzing coastal change over time. The primary advantage of community-based research is that citizen scientists have an invested interest in the future viability of their communities. Once they analyze the collected data and the public sees the results, volunteers are more likely to continue providing useful location-based temporal data, working together to improve the success of scientific and community-based endeavors.

Funded by National Science Foundation Grant #1560015.

References: Brandt, K., Graham, L., Hawthorne, T., Jeanty, J., Burkholder, B., Munisteri, C., & Visaggi, C. "Integrating sketch mapping and hot spot analysis to enhance capacity for community-level flood and disaster risk management." *The Geographical Journal*, 10, 2019.

Drone mission plan for Hopkins Village, Belize

- Image centers
- Flight paths

Structure vulnerability
High
Moderate
Low

Road condition
Very low quality
Low quality
Average quality
High quality
Very high quality
Unknown

Culvert condition
Broken
Damaged
Good

STEPPING UP WITH VOXELS
MODELING SOILS BENEATH THE NETHERLANDS IN 3D

Chris Andrews, Esri

Geologists, geophysicists, petroleum explorers, and mining experts have known something for years that we have more recently acknowledged in the GIS world. To understand the real world at high accuracy, you must experience and explore it in 3D. Our oceans, atmosphere, and the planet underneath us are rich with diverse volumetric data such as rock or soil types, chemical composition, pollutants, noise, aquifers, and even the distribution of life. Oceanographers, climatologists, and geologists have collected data using remote sensing techniques that enabled reconstruction of complex 3D models of real-world systems. Historically, these datasets have often been limited to the purview of scientists or highly technical professionals with specialist software tooling to allow them to explore these data.

With the explosion of interest in and capability for 3D in GIS, users have asked for better capability to view volumetric data in their everyday GIS tools. Our users know that access to volumetric data about the world around them can provide higher-accuracy analysis and better understanding of conditions that they can't physically experience. Access to 3D data in a GIS allows users to easily communicate with non-specialist stakeholders and even enables new types of analyses and workflows that they cannot do with traditional 2D GIS. Using 3D in GIS eliminates the need to use complicated recipes and multiple tools to migrate data from the geophysical, marine, or atmospheric world in GIS experiences or to be used with other GIS content.

Many of the data sources and collection techniques for volumetric data are discrete or discontinuous, resulting in data that may be sparsely distributed throughout a physical space. Techniques exist for filling in, or interpolating, gaps in the volume to enable scientists and engineers to infer the characteristics of any 3D point within the volume. ArcGIS includes a geoprocessing tool for one such technique, called *3D empirical Bayesian kriging*.

In 2D, a cell in a grid of raster data is referred to as a *pixel*. In 3D, we can group interpolated regions into a 3D raster grid. We refer to the cells in this grid as *volumetric elements*, or *voxels*. Voxelization techniques can generate extremely large datasets that are difficult or impossible to view in traditional GIS applications. Academic institutions, petroleum exploration companies, and scientific organizations typically use highly specialized software and hardware systems to view massive voxel datasets. Groups with casual interest in the content, and even less-specialized stakeholders in the same company or organization, often cannot use these expert applications.

ArcGIS can consume volumetric content derived from scientific analysis and remote sensing technology and allow users to display that content alongside any other GIS data. In the ArcGIS workflow, users can read specific types of georeferenced volumetric information, and ArcGIS Pro will convert that data into a "voxel layer" that they can view in a standard ArcGIS Pro 3D scene. Voxels often have a pixelated

or steplike appearance, but users can symbolize them to appear as more analog volumes or continuous gradients.

By consuming voxel data in a GIS, users can combine voxel layers with other standard GIS data types for visualization, exploration, and analysis. Innumerable examples illustrate the use of volumetric information. Engineers and architects see the potential to have rich volumetric information for soils and rocks in the subsurface under existing or proposed construction. Cities can use volumetric information to examine subsurface information and aboveground conditions such as airflow, the effects of heat islands, noise propagation, and aerial pollutants. Marine scientists work in an inherently 3D volumetric space and need better visualization and analysis tools to explore ocean temperature and salinity, freshwater mixing, and the propagation of life throughout the oceans. Even tiny creatures such as plankton occupy massive volumes of water, and ocean currents control their dispersion and aggregation, driven by convection, lunar gravity, and other forces operating on a global scale.

Users should be aware that access to volumetric data can still be inconsistent. In some cases, data simply haven't been collected or created. In other cases, such as in competitive extractive industries, data may be proprietary or protected. However, many government and academic agencies have started sharing volumetric data that may become increasingly useful as more users consume them along with other geospatial content.

NASA, for example, shares large amounts of atmospheric data from satellite studies of Earth. The Dutch independent research organization, TNO (Netherlands Organization for Applied Scientific Research), aggregates and shares massive amounts of subsurface information for use by academia and industry throughout the Netherlands. TNO has been instrumental in working with Esri to help push the limits of what can be done in GIS software.

ArcGIS applications and data types are being used for more comprehensive visualization, exploration, and analysis of 3D content of all types. ArcGIS can combine point clouds, 3D building models, engineering data, and more traditional GIS content. Volumetric data are becoming increasingly relevant in GIS-focused industries. The engineering and construction market is demanding more accurate context for future development to sustain human population growth and to protect the environment. Scientific agencies require more accurate 3D maps of the oceans and atmosphere to combat climate change. Mineral and energy companies use GIS and 3D data to improve target exploration with less environmental impact. Voxel data layers and interactive tools introduce more dynamic, immersive 3D experiences for users to explore, interact with, and analyze the world around them.